NEW KEY GEOGRA

Interactions

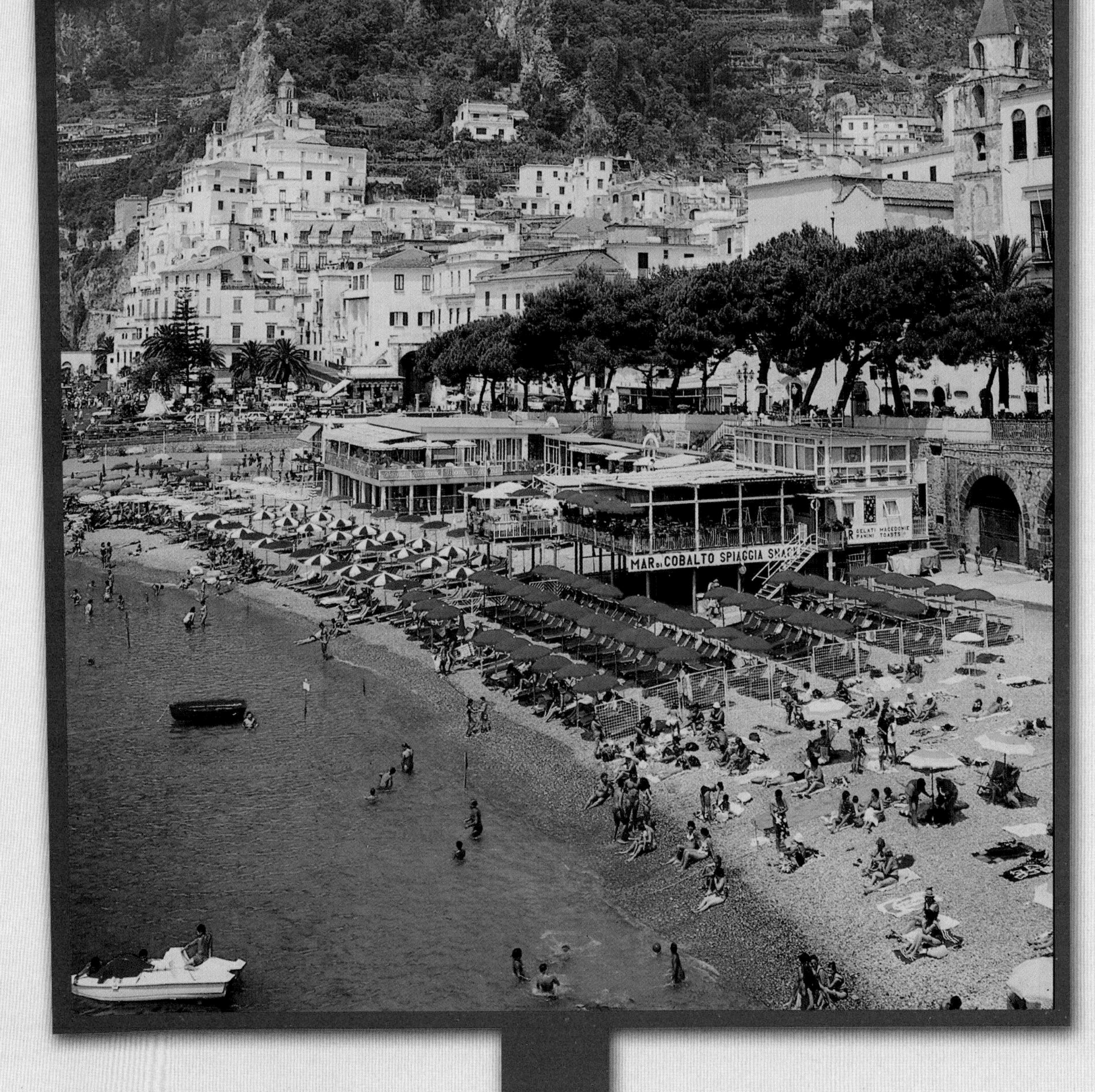

OXFORD
UNIVERSITY PRESS

DAVID WAUGH AND TONY BUSHELL

UNIVERSITY PRESS

Great Clarendon Street, Oxford, OX2 6DP, United Kingdom

Oxford University Press is a department of the University of Oxford. It furthers the University's objective of excellence in research, scholarship, and education by publishing worldwide. Oxford is a registered trade mark of Oxford University Press in the UK and in certain other countries

Text © David Waugh, Tony Bushell 2014
Original illustrations © Oxford University Press 2014

The moral rights of the authors have been asserted

First published by Nelson Thornes Ltd in 2006
This edition published by Oxford University Press in 2014

All rights reserved. No part of this publication may be reproduced, stored in a retrieval system, or transmitted, in any form or by any means, without the prior permission in writing of Oxford University Press, or as expressly permitted by law, by licence or under terms agreed with the appropriate reprographics rights organization. Enquiries concerning reproduction outside the scope of the above should be sent to the Rights Department, Oxford University Press, at the address above.

You must not circulate this work in any other form and you must impose this same condition on any acquirer

British Library Cataloguing in Publication Data
Data available

978-0-7487-9703-5

20 19 18 17 16 15 14

Printed in China by Sheck Wah Tong Printing Press Ltd

Acknowledgements

Cover photos: Digital Vision 14 (NT): left; Narendra Shrestha/ EPA/ epa/ Corbis (middle); Digital Vision XA (NT): right. Title page: Corel 562 (NT).
Illustrations: Kathy Baxendale, Jane Cope, Mike Gordon, Hardlines, Gordon Lawson, Angela Lumley, Richard Morris, David Russell Illustration, Tim Smith
Page make-up: Viners Wood Associates

Photographs
Action Images/ John Sibley: 69D; Actionplus/ Glyn Kirk: 70C (London 2012); Alamy/ Alan King: 70B (Timberland); Alamy/ Barry Lewis: 109C; Alamy/ Eric James: 59D; Alamy/ Eye35.com: 82A; Alamy/ FAN Travelstock: 61B (top); Alamy/ Jack Sullivan: 69C; Alamy/ Jon Arnold Images: 62B & C; Alamy/ Marco Regalia: 98A; Alamy/ Photolibrary Wales: 83C; Alamy/ SCPhotos: 61B (bottom); Alamy/ Stockshot: 57D (bottom right); Art Directors & Trip Photo Library/ Helene Rogers: 77B; Art Directors & Trip Photo Library/ Price: 11 (top right), 104C, 114A, B, C & E; Penni Bickle: 11 (top left); Bluesky International Limited: 80B; Tony Bushell: 33E, 70C (Monaco GP), 76A (top); Val Corbett: 67 (bottom); Corbis/ Bettmann: 26A; Corbis/ Dennis Degnan: 94B; Corbis/ Douglas Peebles: 27C, 46A; Corbis/ Free Agents Ltd: 96 (4); Corbis/ George D Lepp: 47C; Corbis/ Gero Breloer: 68B; Corbis/ Guenter Rossenbach/ Zefa: 97 (7); Corbis/ J Raga/ Zefa: 86B; Corbis/ John Van Hasselt/ Sygma: 116B; Corbis/ Michael S Yamashita: 27B; Corbis/ Owen Franken: 93D (bottom right); Corbis/ Reuters: 35C, 72A (right); Corbis/ Reuters/ London 2012: 80A; Corbis/ Roger Ressmeyer: 44A; Corbis/ Tom Wagner: 132C; Corbis/ Wolfgang Kaehler: 74 (left); Corel 108 (NT): 108A; Corel 684 (NT): 87B; The Countryside Commission: 52B, 53C; Ecoscene/ Stuart Baines: 47D; Empics: 92C; Empics/ AP: 37D, 41C, 42A, 43C, 72A (middle), 100B; Empics/ AP/ Don Ryan: 72A (left); Empics/ DPA: 70C (World Cup 2006); Empics/ PA/ Fiona Hanson: 76A (bottom right); Eye Ubiquitous: 102B, 103C; Eye Ubiquitous/ Anna Barry: 93D (top left); Eye Ubiquitous/ Mike Southern: 98B; GeoScience Features Picture Library: 34B; Getty Images/ Jean-Marc Giboux: 124A; Getty Images/ Manoj Shah: 57D (top right); Getty Images/ Robert Everts: 96 (1); Getty Images/ Stuart Westmorland: 5D; Getty Images/ Will & Deni McIntyre: 5B; Hutchison Picture Library: 20B; Hutchison Picture Library/ Edward Parker: 50A; Hutchison Picture Library/ Richard House: 132B; Hutchison Picture Library/ Robert Francis: 93D (top right); ICCE Photolibrary/ John Parrott: 15B; James Davis Travel Photography: 11 (top middle, bottom left & right), 92B, 99E, 114D; Japan National Tourist Board: 111D; John Warburton-Lee Photography: 125D; Lonely Planet Images/ Andrew Lubran: 75E (left); Lonely Planet Images/ Diana Mayfield: 87C; Lonely Planet Images/ Ray Laskowitz: 68A; Magnum Photographers/ Berry: 115F; Mazda Corporation: 117D; National Trust Picture Library/ Joe Cornish: 67 (top); Nature Picture Library/ Aflo: 4A; Nature Picture Library/ Peter Oxford: 125B; Network Photographers/ Lewis: 99D; Panos/ Fernando Moleres: 74B; Panos/ Jeremy Hartley: 137D; Panos/ Qilai Shen: 74A; Photofusion/ Molly Cooper: back cover (left); Photolibrary/ Jon Arnold Images: 47B, 104B; Reportdigital.co.uk/ Philip Wolmuth: back cover (right); Reuters: 38B, 39 (both); Rex Features/ Alex Segre: 70B (Gap); Rex Features/ Jennifer Jacquemart: 70B (Reebok); Rex Features/ Roy Garner: 109D; Rex Features/ Sipa Press: 83B, 109B; Rex Features/ Sten Rosenlund: 40B; Robert Harding Photo Library: 13E, 19C, 93D (bottom left); Robert Harding Photo Library/ Paul van Riel: 118C; Robert Harding Photo Library/ Walter Rawlings: 97 (2); Science Photo Library/ David Parker: 40A; Science Photo Library/ Earth Satellite Corporation: 118 (right); Science Photo Library/ NRSC Limited: 95C; Science Photo Library/ Peter Menzel: 36C; Still Pictures: 11 (bottom middle), 12B, 20C, 22C, 75E (right); Still Pictures/ Charlotte Thege: 5C; Still Pictures/ Ron Giling: 75C, 125C; Swift Imagery: 111E; The Travel Library Limited: 57D (top left); David Waugh: 19B; World Pictures: 16, 50B, 59C, 96 (3).

Maps produced from Ordnance Survey mapping with the permission of Ordnance Survey on behalf of HMSO. © Crown copyright (2006). All rights reserved. Ordnance Survey Licence number 100036771: 81C (Landranger 139), 81D (Landranger 176), back cover (Explorer 162 & 174).

With thanks to Solo Syndication for permission to reproduce material from the article 'The whole world was shaking' (36C), *Daily Mail*, 9.10.1989.

Although we have made every effort to trace and contact all copyright holders before publication this has not been possible in all cases. If notified, the publisher will rectify any errors or omissions at the earliest opportunity.

Links to third party websites are provided by Oxford in good faith and for information only. Oxford disclaims any responsibility for the materials contained in any third party website referenced in this work.

The previous page sho[w]
the coastal town of Ama[l]
in the Compania region,
Southern Italy

Contents

1 Ecosystems

How are ecosystems at risk?

What is this unit about?

This unit looks at different types of ecosystems. It explains how their characteristics and distribution can be linked to factors such as climate and soils. It also looks at how ecosystems may be changed by human activity.

In this unit you will learn about:

- **factors that affect climate**
- **the British, equatorial and Mediterranean climates**
- **the main features of ecosystems**
- **the characteristics of tropical rainforest and Mediterranean ecosystems**
- **the causes and effects of soil erosion.**

Why is this an important topic?

Learning about ecosystems will help you understand our world and appreciate the different types of vegetation and wildlife found across the earth's surface. It will also help you understand why we need to use the earth's resources carefully so that we do not damage or change the environment for future generations.

Learning about ecosystems will help you to:

- appreciate the world about you
- develop a concern for the environment
- learn about protecting wildlife and scenery
- develop an interest in your surroundings.

A Deforestation in Brazil

B The Amazon rainforest

- What is happening in photo **A** and what problems may this cause?
- What has happened in photo **C**?
- Describe the vegetation and wildlife in photos **B** and **D**. In what ways are they different from where you live?

C Soil erosion in Kenya

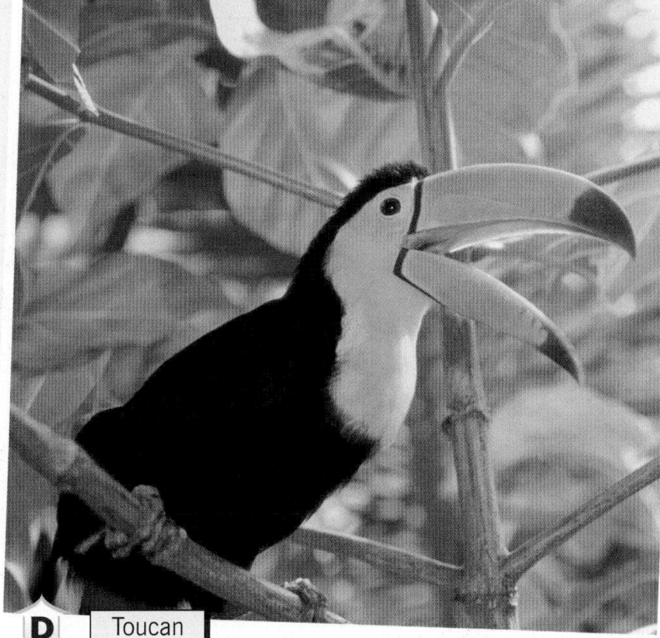

D Toucan

What factors affect climate?

Can you remember the difference between weather and climate shown on diagram **A**? There are several types of climate found across the world. Each type has its own distinctive pattern of temperature and rainfall.

Several factors affect climate. Four of these factors are **latitude**, the **distance from the sea**, **prevailing winds** and major **relief** features. It is important to understand these factors before looking at Britain's climate and comparing it with other places in the world.

A

B

Latitude

Places near the Equator are hotter than places near the poles. This is due mainly to the curvature of the earth and the angle of the sun.

At the Equator the sun is often overhead. It shines straight down and its heat is concentrated on a small area which gets very hot.

Towards the poles the sun shines more at an angle. Its heat is spread over a larger area and temperatures are lower.

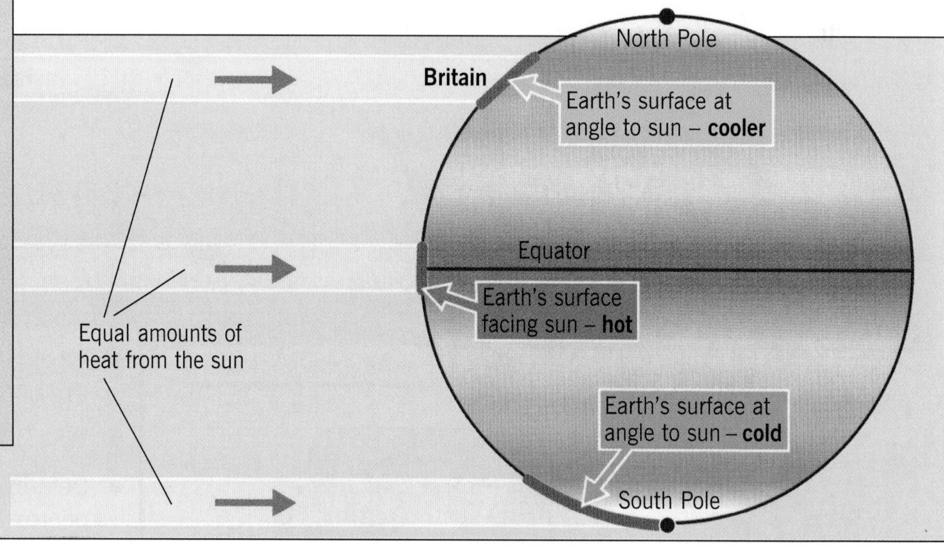

C

Distance from the sea

The distance a place is from the sea affects its temperature. In summer places which are inland and away from the sea are usually warmer than places near to the coast. In winter it is usually the opposite with inland places being cooler than places near to the coast.

Imagine you have to do the washing-up at home and there is no hot water. You fill a kettle with cold water and have to wait for a few **minutes** for it to heat up. You put the hot water and some cutlery in a bowl. Within **seconds** the cutlery becomes hot. After washing, the cutlery cools down in a few seconds but the water in the bowl stays warm for much longer. This is because liquids, like the sea, take longer to heat up than solids like the land. Once warm, liquids keep their heat for much longer than solids.

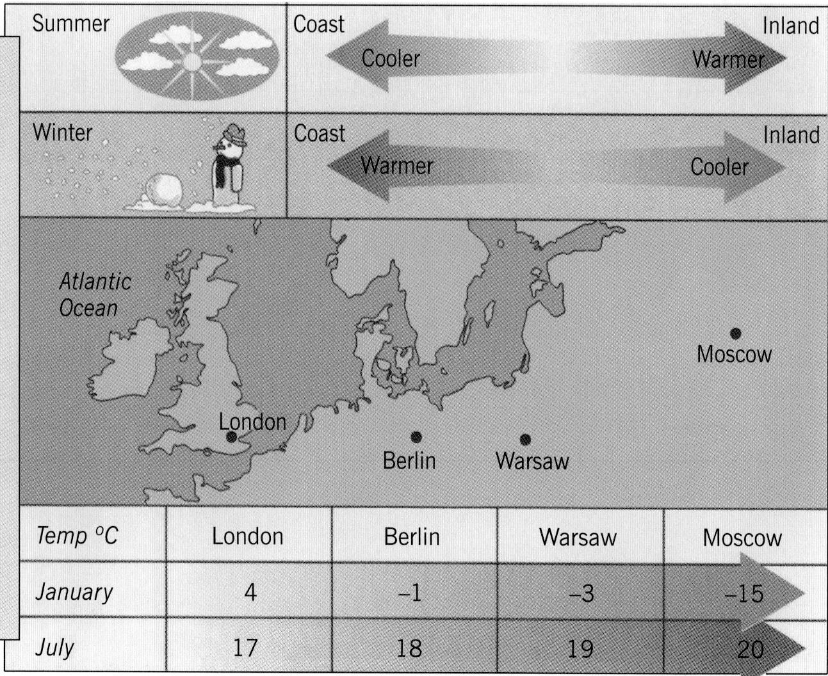

Temp °C	London	Berlin	Warsaw	Moscow
January	4	–1	–3	–15
July	17	18	19	20

Prevailing winds

The prevailing wind is the direction from which the wind usually comes. In Britain the prevailing wind is from the south-west.

Prevailing winds will bring:

- cool weather if they blow over cool surfaces such as the land in winter or the sea in summer
- warm weather if they blow over warm surfaces such as the sea in winter or the land in summer
- wet weather if they blow over sea areas and pick up moisture
- dry weather if they blow over the land.

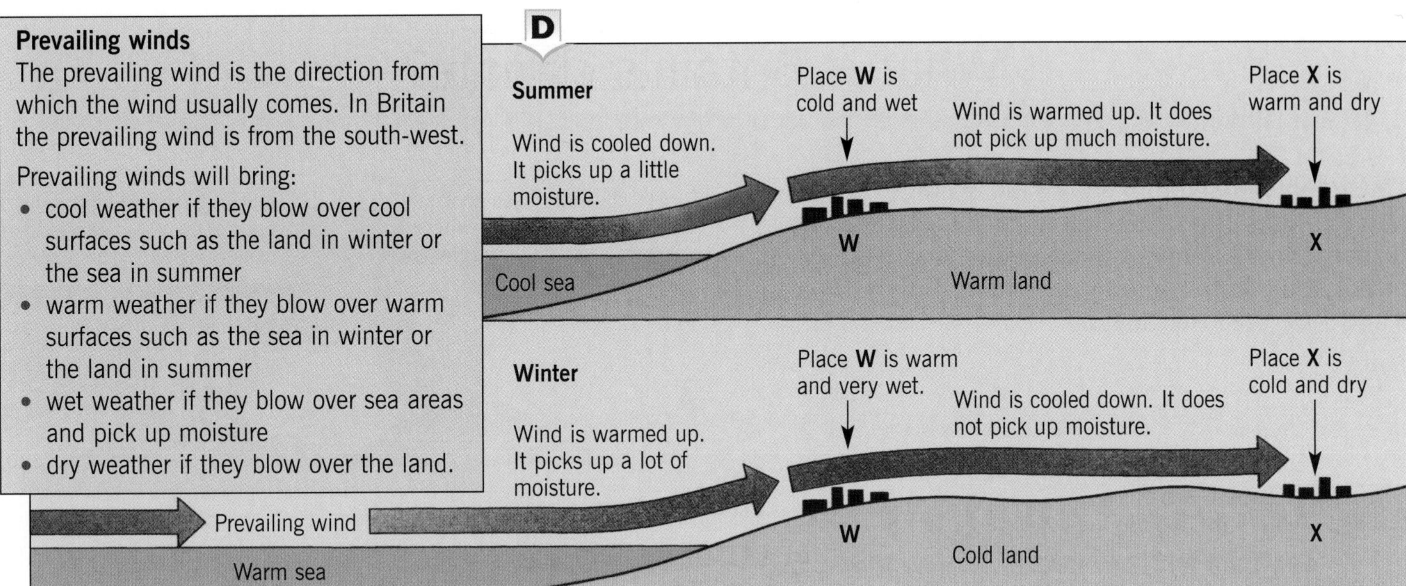

Relief (altitude)

Places which are high up and in mountains have lower temperatures and more rainfall than places which are lower down.

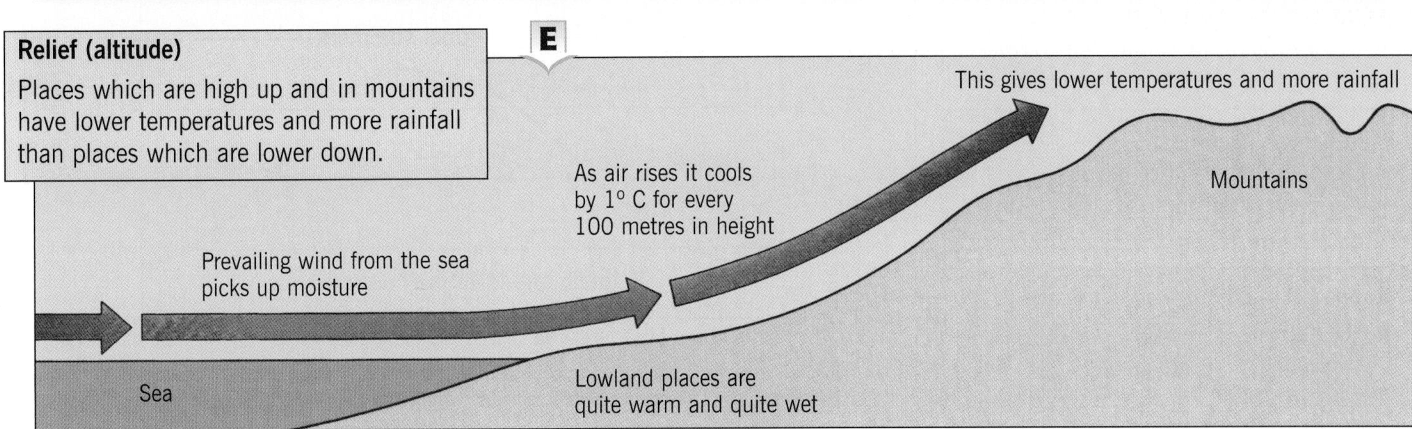

Activities

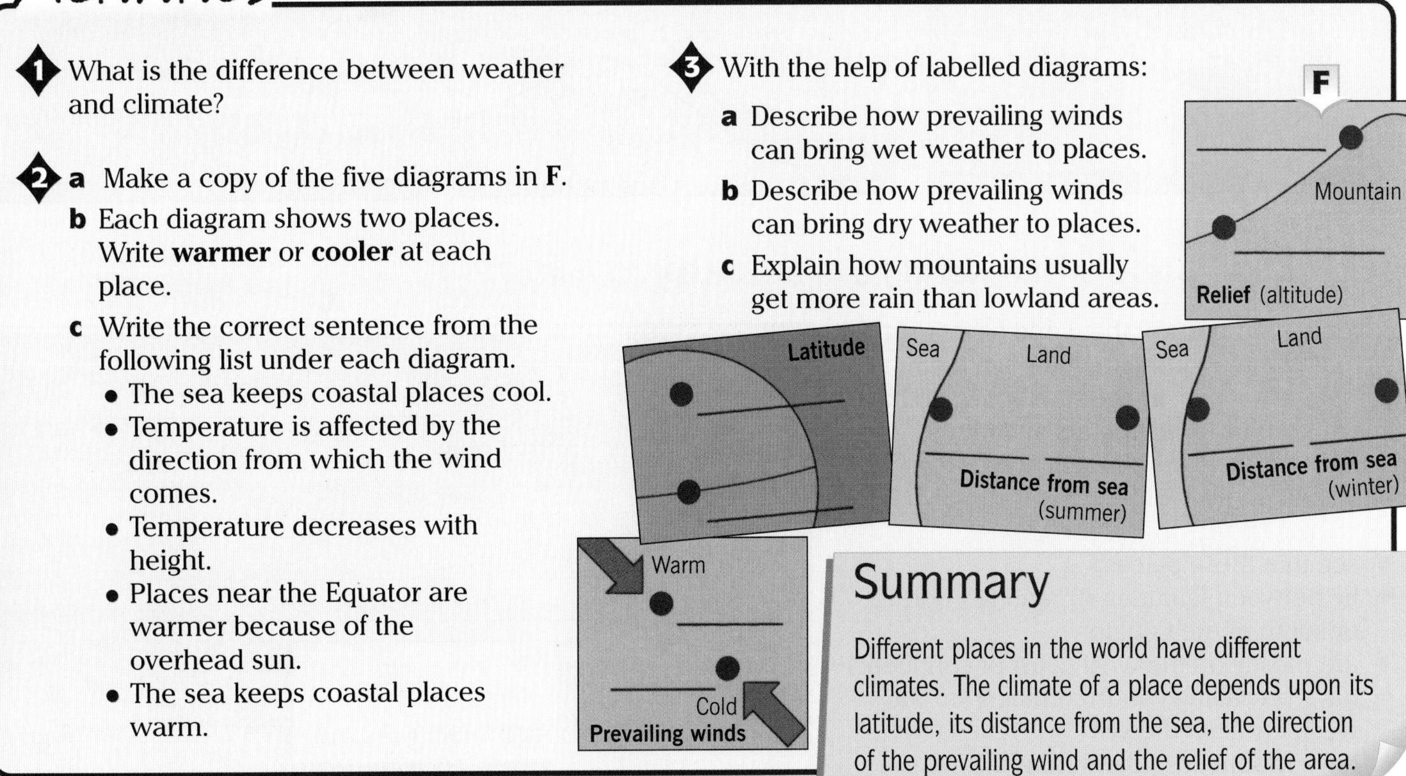

1 What is the difference between weather and climate?

2 **a** Make a copy of the five diagrams in **F**.

b Each diagram shows two places. Write **warmer** or **cooler** at each place.

c Write the correct sentence from the following list under each diagram.

- The sea keeps coastal places cool.
- Temperature is affected by the direction from which the wind comes.
- Temperature decreases with height.
- Places near the Equator are warmer because of the overhead sun.
- The sea keeps coastal places warm.

3 With the help of labelled diagrams:

a Describe how prevailing winds can bring wet weather to places.

b Describe how prevailing winds can bring dry weather to places.

c Explain how mountains usually get more rain than lowland areas.

Summary

Different places in the world have different climates. The climate of a place depends upon its latitude, its distance from the sea, the direction of the prevailing wind and the relief of the area.

What is Britain's climate?

The graph in diagram **B** shows the average monthly temperatures and rainfall for a place in Britain. The reasons why Britain has this type of climate are given beside the graph. You may have to turn back to pages 6 and 7 for their explanation.

A

Britain usually has cool summers and mild winters.
Although rain can fall at any time of the year, winters are usually wetter than summers.

B

Cool summers are due to:
- Britain's latitude (50–60° north) being a long way from the Equator
- the cooling effect of the sea
- the prevailing wind from the cool sea.

Mild winters are due to:
- the warming effect of the sea
- the prevailing wind from the warm sea.

Rainfall throughout the year is due to the prevailing wind picking up moisture as it crosses the sea. Three types of rain which can fall in Britain are shown below.

Temperature (°C): 0, 5, 10, 15, 20, 25
Rainfall (mm): 0, 40, 80, 120, 160, 200
Months: J F M A M J J A S O N D
Cool summers
Mild winters
Total rainfall 900 mm
Rainfall spread throughout the year

Frontal rain falls when warm air has to rise over cold air.

Warm air rises
Cold air

Relief rain falls when warm, moist air is forced to rise over mountains.

Mountains
Sea

Convectional rain (mainly summer) occurs when heated ground warms the air next to it. The warm air then rises, causing thunderstorms.

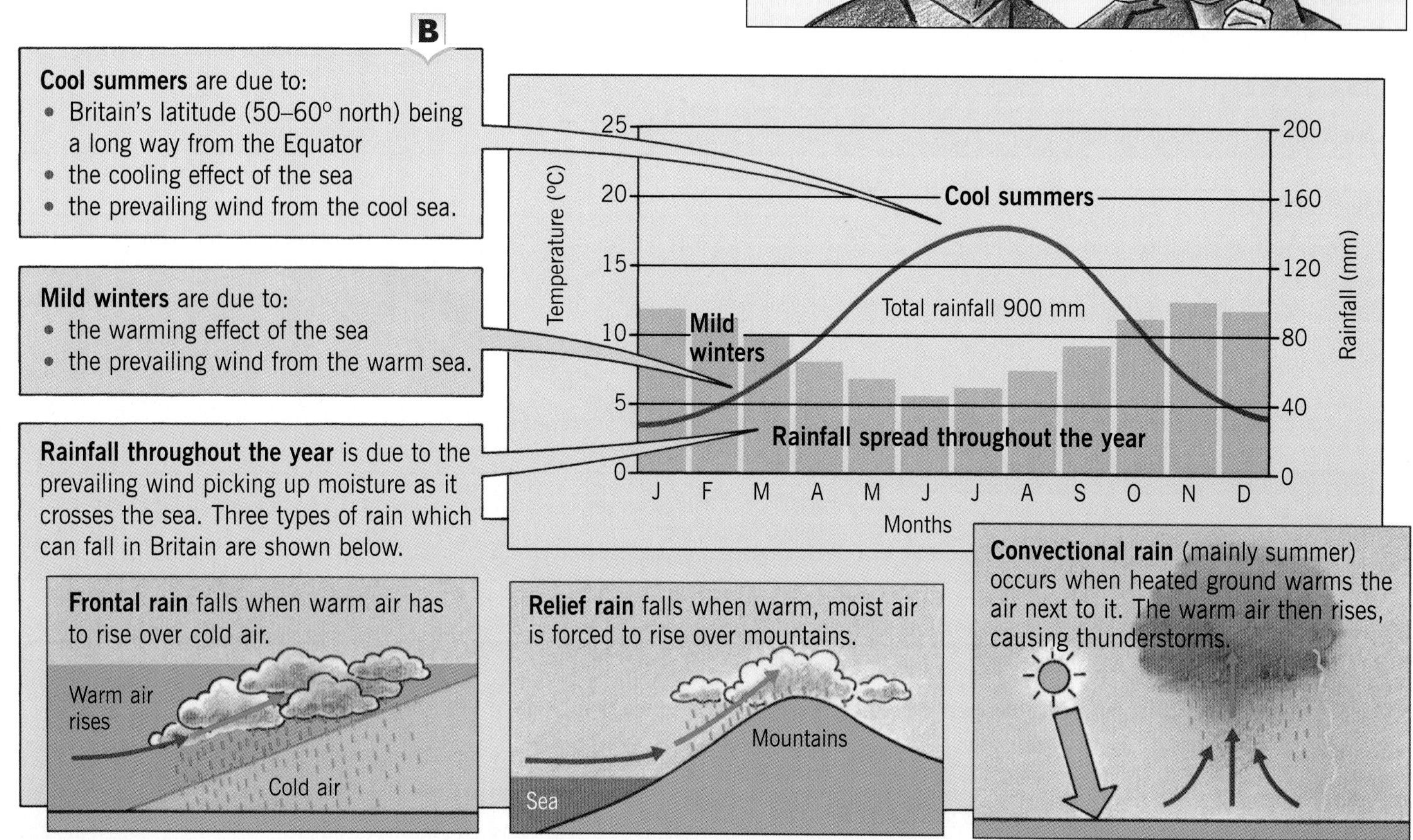

What other parts of the world have a 'British' climate?

Several other parts of the world have a similar type of climate to Britain. Although their climate is not exactly the same, they do have cool summers, mild winters and rain throughout the year. These places are shown on map **C**.

Notice that these places:
- lie between latitudes 40° and 60° north or south of the Equator
- are mostly on the west coast of continents
- have prevailing winds coming from the west, i.e. from the sea
- have mountains inland from the coast.

C

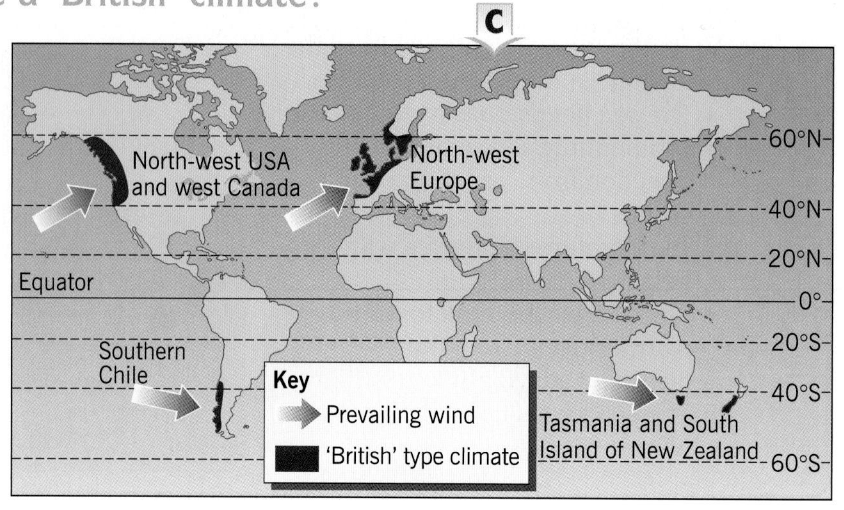

How does the climate differ in other parts of the world?

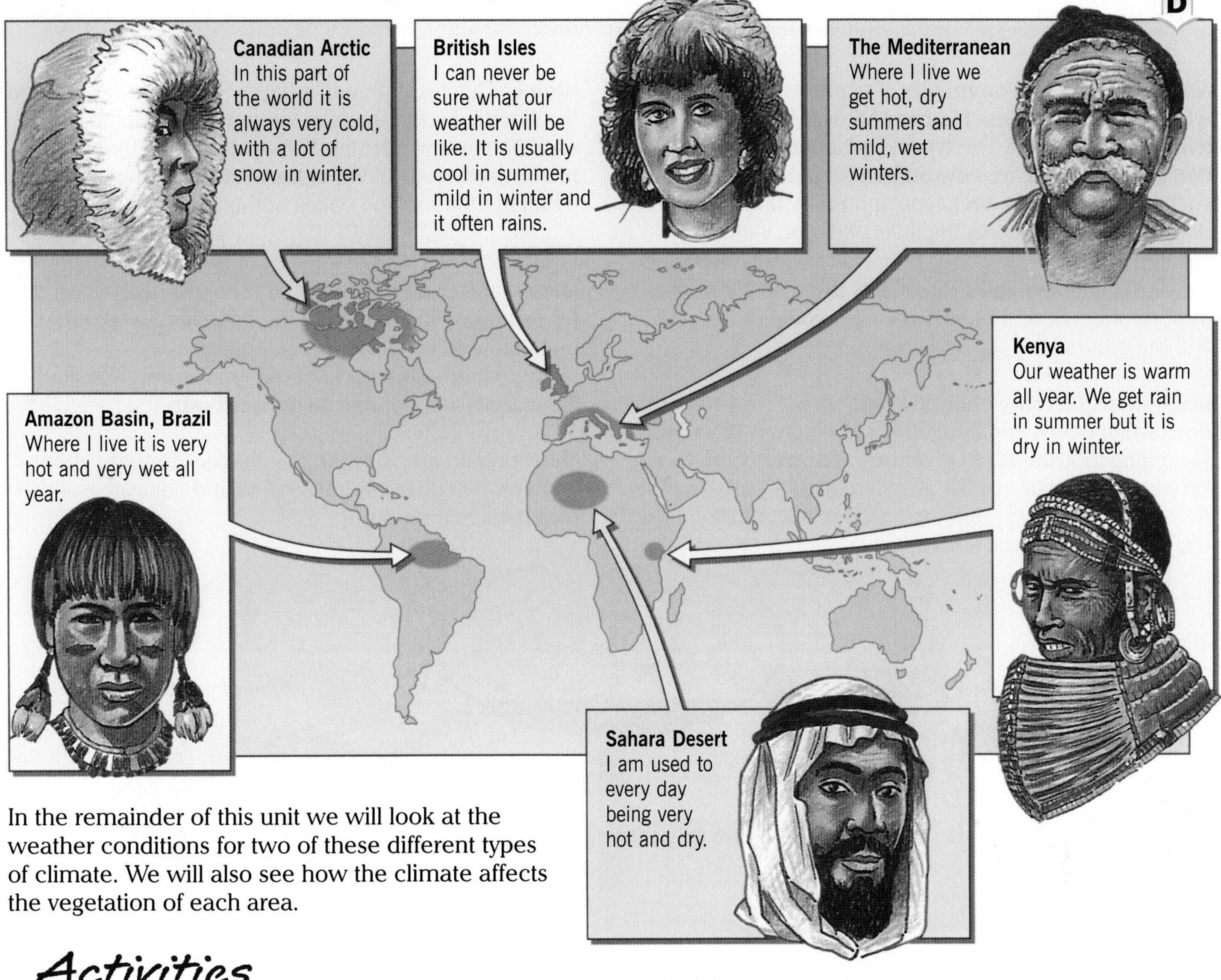

In the remainder of this unit we will look at the weather conditions for two of these different types of climate. We will also see how the climate affects the vegetation of each area.

Activities

1. Copy out and complete the sentence below:

   The British Isles have __________ summers, __________ winters and __________ throughout the year.

2. Name four places in the world which have the same type of climate as Britain.

3. Which **four** of the following statements are correct about the British climate?
   - Summers are cool because Britain is a long way from the Equator.
   - Summers are hot because the sun is overhead.
   - Winters are mild due to warm winds blowing from the sea.
   - Winters are mild because Britain is near to the Equator.
   - Rain falls all year because winds blow from the sea.
   - Britain gets frontal and relief rainfall all year, and convectional rainfall in summer.

Summary

The British climate has mild winters and cool summers, and rain falls throughout the year.

What are ecosystems?

An **ecosystem** is a community of plants and animals which interact with each other and with their **non-living environment**. The types of plants and animals that grow or live in an environment depend on factors such as climate, soil, rock type and relief. Look at diagram **A** which shows the links between various elements of a simple ecosystem. Notice how the parts are closely related and each affects the others. This is important because if one component changes, then so will the others.

Ecosystems can vary enormously in scale. They range from small ponds and hedgerows to whole **rainforests** stretching thousands of kilometres across continents. A very large ecosystem such as a rainforest is called a **biome**. The vegetation in a biome is largely determined by climatic factors such as temperature, rainfall, sunshine and wind. Map **B** shows some of the world's major ecosystems. Two of these are explained in more detail in the next few pages of this unit.

In recent times, ecosystems have been increasingly altered and even destroyed by human activity. In Brazil, for example, 40 per cent of the original rainforest has been cleared for development, whilst across Scandinavia huge areas of coniferous forest are slowly dying from air pollution called **acid rain**.

Many people are worried that changes brought about by human activity are irreversible and will cause severe problems in the future.

A Links in a simple ecosystem

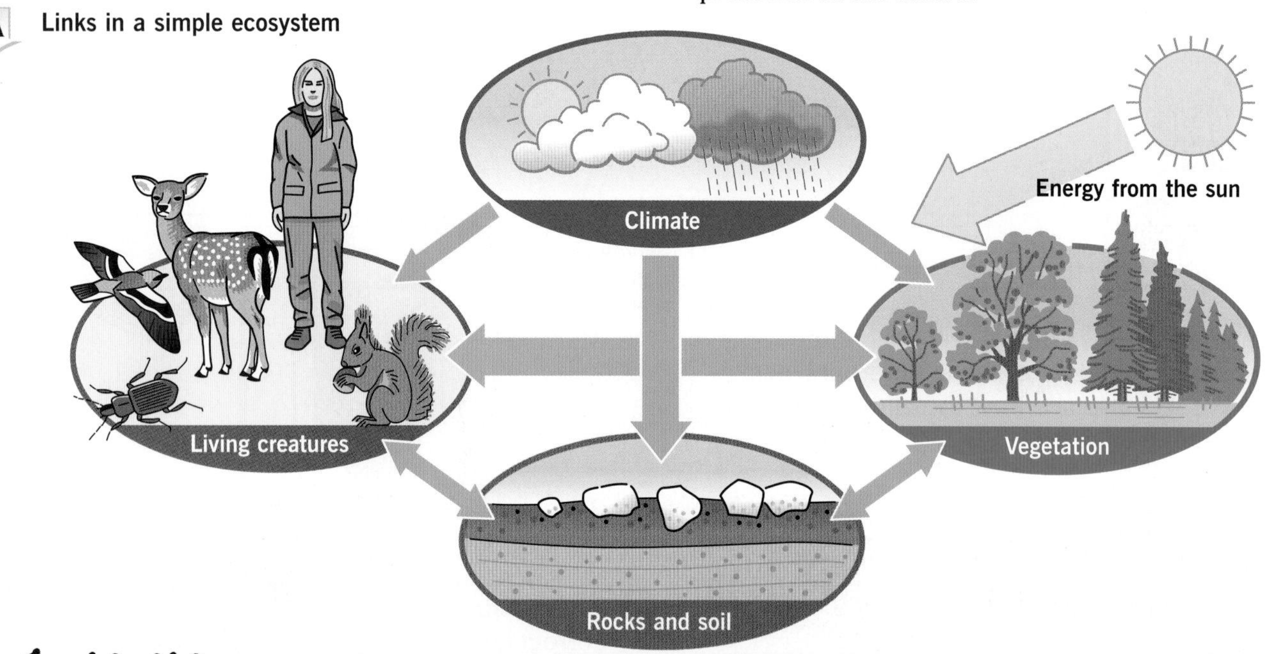

Activities

Complete the following sentences using the descriptions in the boxes.

a The environment is ...
b The non-living environment includes ...
c The living environment includes ...
d An ecosystem is ...
e A natural ecosystem is ...
f A biome is ...

... the interaction of plants and animals with their non-living surroundings.

... a large ecosystem such as a desert or tropical rainforest.

... animals, birds, fish, insects and people.

... rocks, soil, the air and climate.

... the natural or physical surroundings where plants and animals live.

... a community of plants and animals unaffected by human activity.

B Some major world ecosystems (biomes)

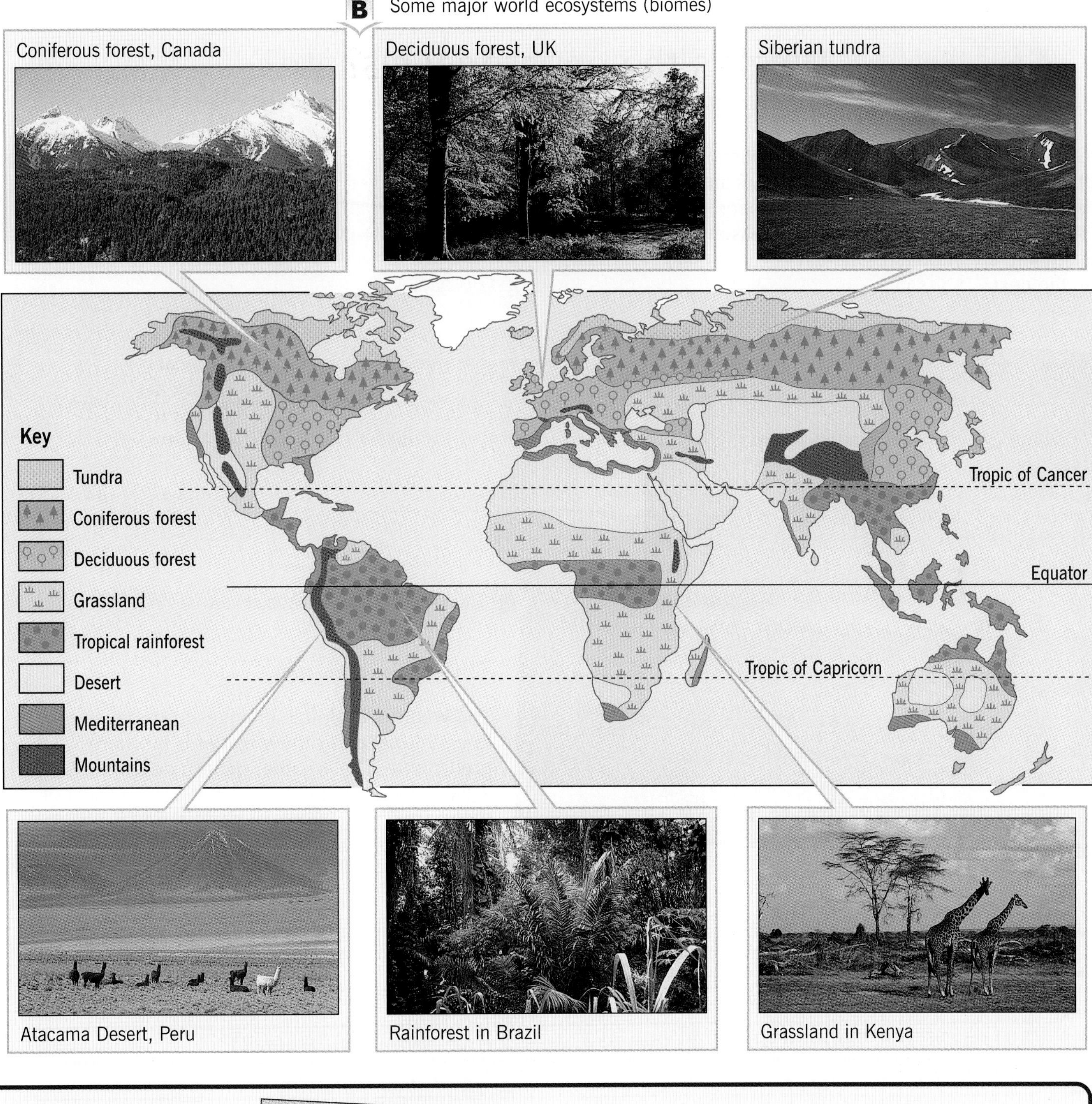

C

Acid rain damage hits Scandinavian forests

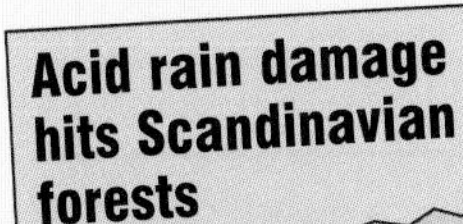

Kenyan wildlife threatened by farmers ploughing and overgrazing the land

Arctic oil spill threatens caribou grazing land

Problems for Brazil as miners and loggers clear more forest

Woodland cleared for Lake District leisure complex

Global warming blamed for spread of desert

2 For each of the headlines shown in **C**:

a name the type of biome it refers to

b suggest how plants, animals or people might benefit or suffer from the changes.

Give reasons for your answers.

Summary

An ecosystem is a community of plants and animals whose lives are closely linked to each other and to the climate and soil of the area in which they grow or live. Ecosystems may be changed by human activity.

What is the equatorial climate?

Graph **A** shows the **equatorial** type of climate. It is hot and wet throughout the year. Rainfall is heavy and falls during most afternoons. There are no winters or summers (seasons) like there are in Britain. One day is very similar to the next.

A

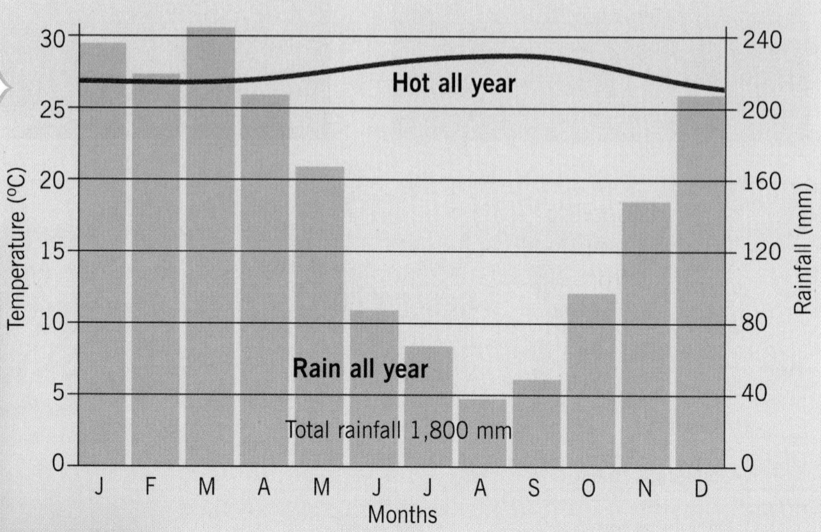

B Thunderstorm above the rainforest

The daily pattern

The weather in Britain changes from day to day. In equatorial areas the weather is far more predictable. The weather pattern described below is likely to be repeated day after day for most of the year.

C

Time	Weather conditions
6.00 a.m. (0600)	Sun rises as always at this time. No clouds in sky.
7.00 a.m. (0700)	Gets warmer. Very little wind.
8.00 a.m. (0800)	Temperature 25°C (same as a warm summer afternoon in England)
9.00 a.m. (0900)	Temperature continues to rise. Becomes very hot as the
10.00 a.m. (1000)	sun gets higher in the sky. Hot air rises. Water
11.00 a.m. (1100)	from rivers, swamps and vegetation evaporates.
Midday	Temperature reaches 33°C. Sun overhead. Hot air continues to rise.
1.00 p.m. (1300)	Water vapour carried high into sky. Cools and condenses to form white <u>cumulus</u> clouds.
2.00 p.m. (1400)	Clouds increase in size and height. Turn into towering dark grey <u>cumulo-nimbus</u> rainclouds.
3.00 p.m. (1500)	Torrential rainstorm with thunder and lightning.
4.00 p.m. (1600)	Storm continues.
5.00 p.m. (1700)	Storm ends. Clouds begin to break up.
6.00 p.m. (1800)	Sun sets. The night is clear and calm.

Which places have an equatorial climate?

Map **D** shows this type of climate to be mainly limited to places within 5° north or south of the Equator. The two main areas where it is found are the huge river basins of the Amazon in South America and the Congo in Africa. The major factor which affects this climate is its latitude. The sun is overhead throughout the year. This gives high temperatures and is responsible for the convectional rainfall. There is no prevailing wind and the air is calm apart from during thunderstorms.

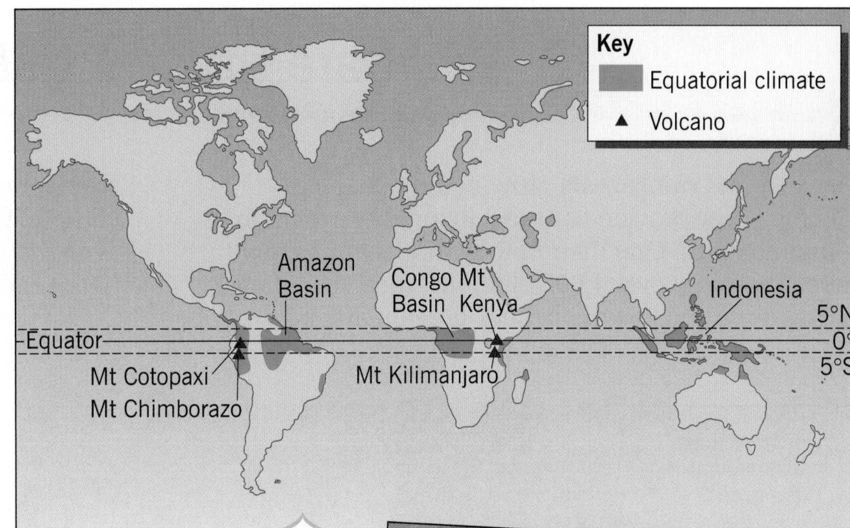

D

How can height affect different places with the same latitude?

However, not all places near to the Equator have this type of climate. This is because of **altitude**. In East Africa two mountains, Kilimanjaro and Kenya, rise to nearly 6,000 metres (19,000 feet). Two South American volcanoes, Cotopaxi and Chimborazo, rise to similar heights in the Andes. As photo **E** of Mount Kenya shows, this makes it cold enough for snow to lie all year on the mountain summits.

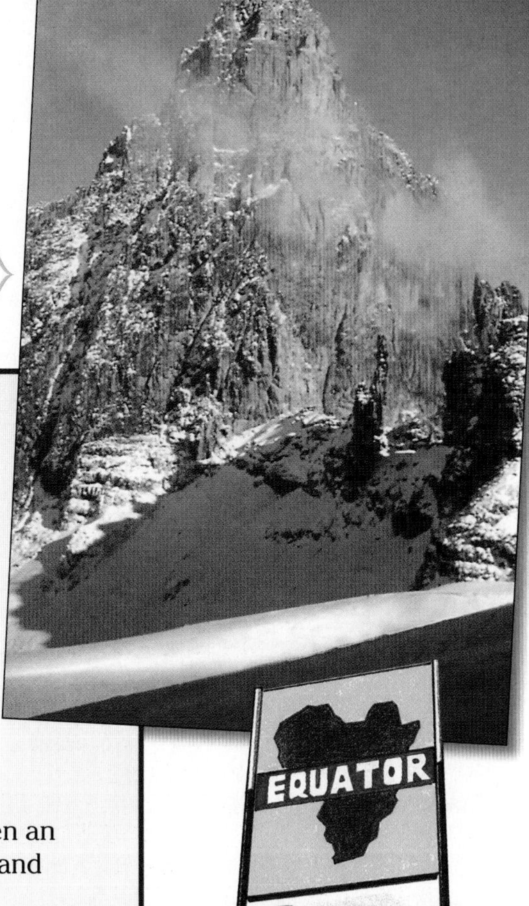

E

Mount Kenya

Activities

1. Write out the paragraph below choosing the correct word from the pair in brackets.

   The equatorial climate is (hot/warm) and (dry/wet) all year. Rainfall is (heavy/light) and falls during most (mornings/afternoons). The weather for each day of the year is very (different/similar).

2. Look at map **D**. Which **two** of the statements below correctly describe the location of the equatorial climate?
   - It is mainly found between latitudes 5°N and 5°S of the Equator.
   - All places on the Equator have this type of climate.
   - It is found in the Amazon Basin in South America and in the Congo Basin in Africa.

3. Using graph **A**:

   a What is the temperature for the equatorial climate in
   - January
   - July?

   b Why is it hot throughout the year?

4. **a** Using graph **A**, how much rain falls in a year?

   b Why do some places near to the Equator have snow lying all the year?

5. Describe three differences between an equatorial climate and Britain's climate.

EQUATOR

W N E S

Summary

Due to its latitude the equatorial climate is hot and wet throughout the year. The weather for one day is very similar to that of the next.

What are tropical rainforests?

The **tropical rainforests** grow in the equatorial climate. They provide the most luxuriant vegetation found on earth. One third of the world's trees grow here. There are over 1,000 different types, including **hardwoods** such as mahogany, greenheart and rosewood. The vegetation has had to **adapt** to the climate. By adapt we mean that it has had to learn to live with the constant high temperatures and the heavy rainfall (diagram **A** and photo **B**).

A

How vegetation has adapted to the equatorial climate

- The trees can grow to over 40 metres in the effort to get sunlight.
- The forest has an **evergreen** appearance due to the continuous growing season. This means that trees can shed leaves at any time, but always look green and in leaf.
- The leaves have drip tips to shed the heavy rainfall.
- Tree trunks are straight and branchless in their lower parts in their efforts to grow tall.
- **Lianas**, which are vine-like plants, use large trees as a support to climb up to the canopy.
- The forest floor is dark and damp. There is little undergrowth because the sunlight cannot reach ground level.
- Dense undergrowth develops near rivers or in forest clearings where sunlight can penetrate.
- Rivers flood for several months each year.
- Fallen leaves soon rot in the hot, wet climate.
- Large **buttress roots** stand above the ground to give support to the trees.

Most of the 1,000 types of trees have yet to be studied. Many may prove to be valuable to us. Over half of our known drugs (e.g. quinine which is used to treat malaria) have come from here. Recently one plant, a type of periwinkle, has been found to be successful in treating leukaemia in children. It is hoped that the rainforests may be a source of cures for other illnesses such as cancer and AIDS.

B Tropical rainforest, Cameroon

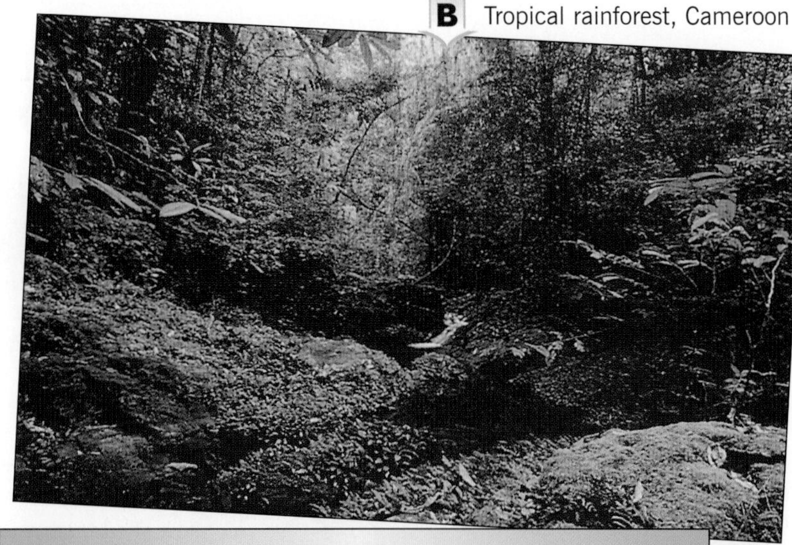

Wildlife

Apart from large animals which would find it difficult to move between the large trees, the rainforest is full of wildlife (diagram **C**). Some **species** live in the **canopy**; others live either in the **under-canopy**, on the forest floor or in the many swamps and rivers. Like the trees, wildlife has had to adapt in order to survive in this climate and vegetation. In many areas of rainforest, wildlife is threatened by **deforestation** and other human activities.

C

A typical 10 kilometre square of rainforest might include:

- 750 species of tree
- 400 species of bird
- 150 species of butterfly
- 1,500 species of flowering plant
- 20 types of animal
- 100 types of reptile
- 60 types of amphibian
- countless numbers of insects and fish.

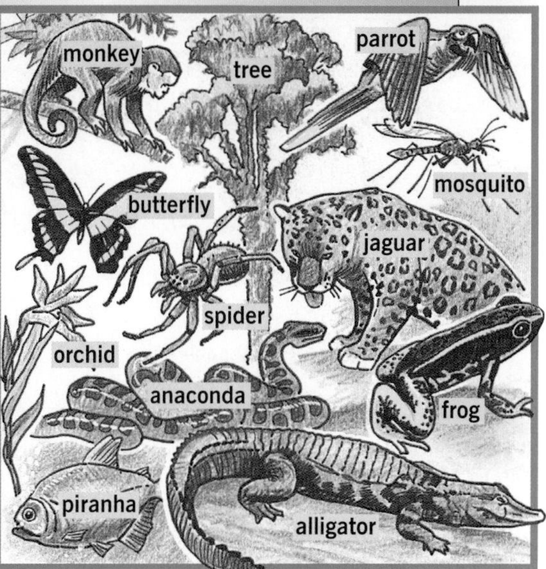

Activities

1. Match up each of the seven descriptive points of the tropical rainforest from the list below with its correct number from diagram **D**.

   buttress roots | main canopy | lianas | under-canopy | branchless trunks | little undergrowth | shrub layer

   Answer like this:

   1 = Little undergrowth

2. **a** Why do trees grow so tall?
   b Why are buttress roots needed?
   c Why is there so little undergrowth?
   d Why do plants grow so quickly?

D

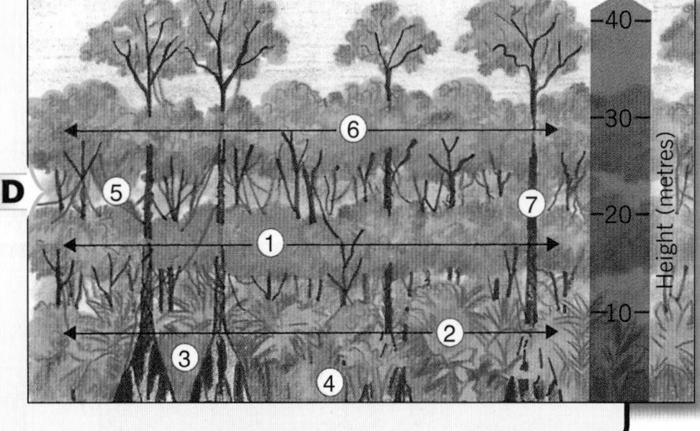

EXTRA

Describe how wildlife in the rainforest has had to adapt to living in the hot, wet, forest environment.

Summary

The tropical rainforests have more trees and wildlife than anywhere else on earth. All plants and wildlife have had to adapt to the hot and wet environment.

What is a Mediterranean climate?

Graph **B** shows the **Mediterranean** type of climate. It has two very different seasons. The weather in summer is hot and dry while in winter it is warm and wet.

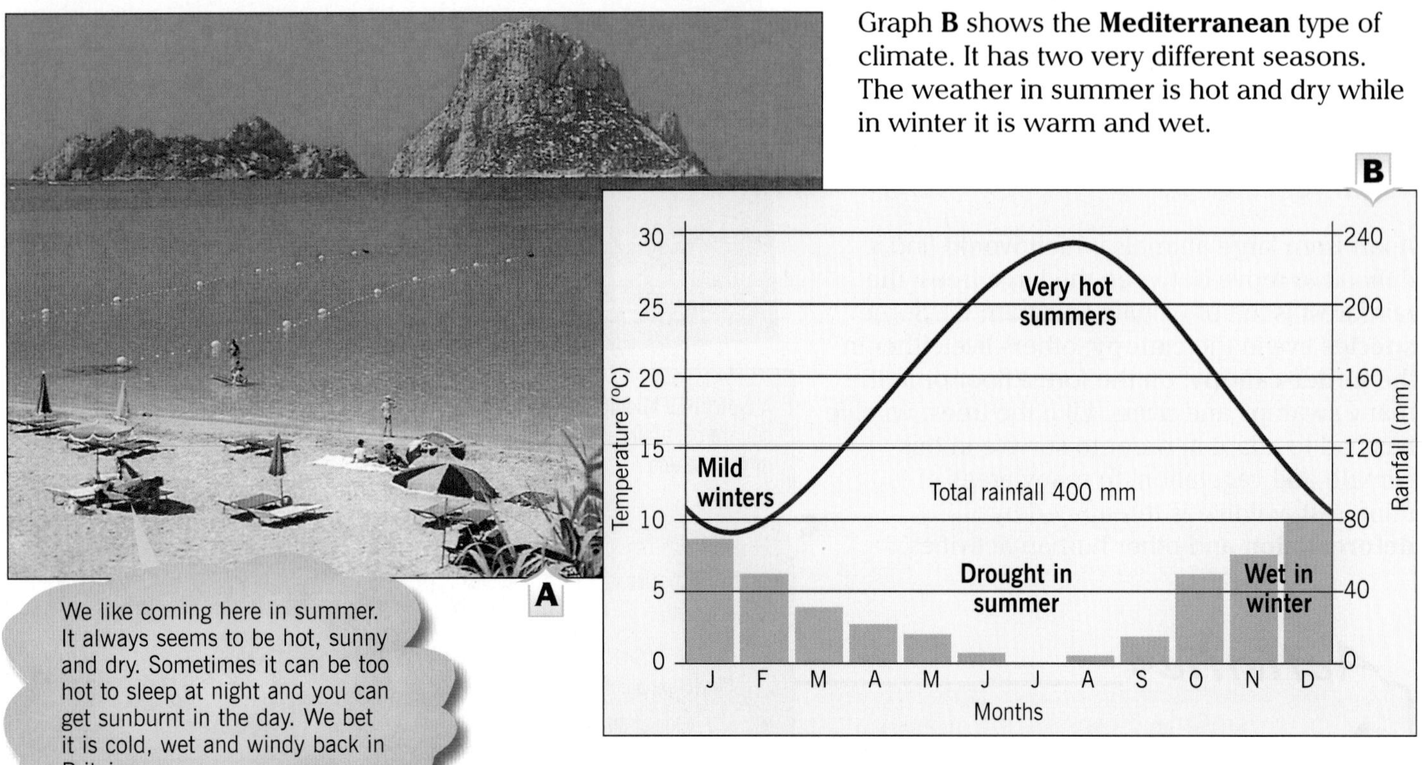

Why are summers hot and dry?

Summers are hot because the sun rises high into the sky. Although it does not shine from directly overhead as it does nearer the Equator, it does rise higher than in places to the north, like Britain. The prevailing wind blows from the land (map **C**). As the land is hot at this time of year then the wind blowing from it will bring hot weather. As the land is also dry then the wind blowing over it cannot pick up much moisture. This means that most places have very little rain and several months of drought. Apart from an occasional thunderstorm most days are cloudless and sunny.

Why are winters warm and wet?

Although the sun is lower in the sky in winter it is still high enough to give warm days. The nearby sea, which was warmed during the summer, only loses its heat slowly in winter. This keeps places near to the coast warm. Frost and snow are unusual near sea-level. The prevailing wind blows from the opposite direction to that of summer (map **D**). As it now comes from the sea it brings air that is warm and moist. As the air rises over the many coastal mountains it gives large amounts of relief rainfall and, at higher altitudes, snow. However, wet days are usually separated by two or three days which are warm and sunny.

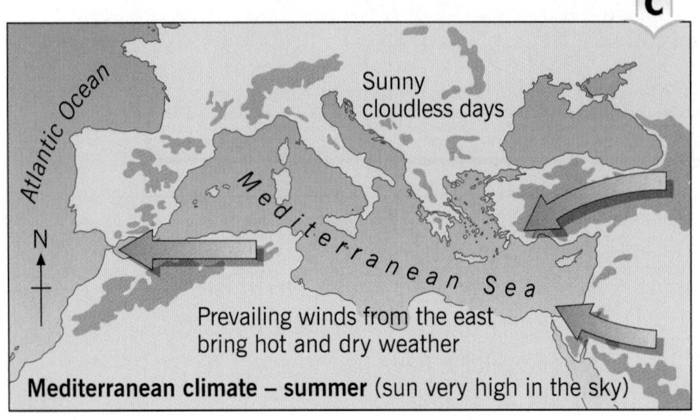

Mediterranean climate – summer (sun very high in the sky)

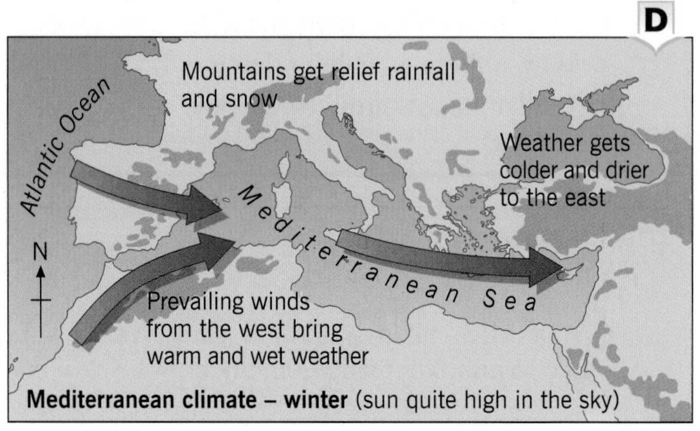

Mediterranean climate – winter (sun quite high in the sky)

Which places have a Mediterranean climate?

The name 'Mediterranean' is given to climates in several different parts of the world. Map **E** shows that this type of climate is usually found:

- on the west coasts of continents
- between latitudes 30° and 40° north and south of the Equator.

In summer the prevailing wind comes from the east and the weather is like that of the hot deserts (map **C**). In winter the prevailing wind comes from the west and the weather is more like ours in Britain (map **D**).

E

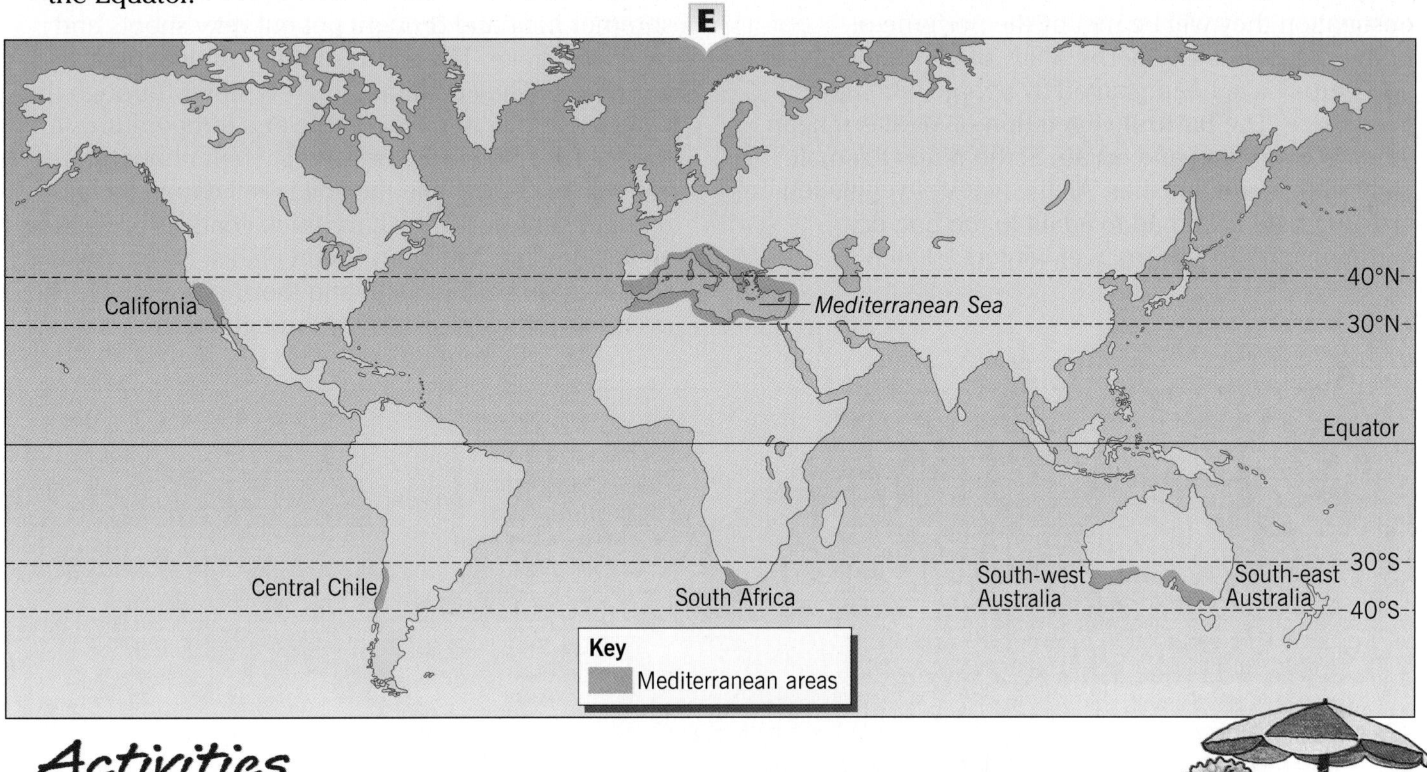

Activities

1 Name the places in the world which have a Mediterranean type of climate.

2 **a** Make a large copy of diagram **F**.

b Complete the sentences by adding the correct endings from the list in the box below:

> ...the prevailing wind blows from the land.
> ...the sun rises high in the sky.
> ...the warming effect of the sea.
> ...the prevailing wind blows from the sea.

F

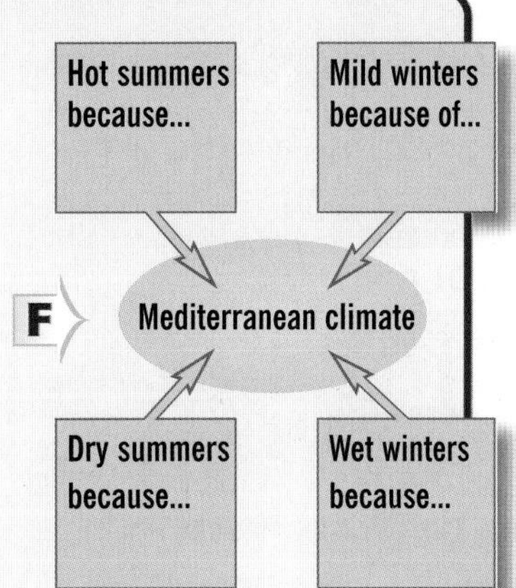

3 **a** What is the temperature for the Mediterranean climate in:
- January
- July?

b How much rain falls in an average year?

EXTRA

Why does the Mediterranean type of climate attract:

a many British families for their summer holidays

b some elderly British people for most of the winter?

Summary

Places with a Mediterranean climate have hot, dry summers and warm, wet winters.

What is Mediterranean vegetation like?

Visitors to the Mediterranean in the spring and summer will see masses of brightly coloured flowers as they arrive by air or road. On reaching their destination they will be met by the perfume of numerous plants and herbs. Many of these flowers and herbs have been planted by people who have lived there. The **natural vegetation** of Mediterranean lands is woodland and **scrub**. Scrub refers to small, stunted trees and bushes. All the types of vegetation growing here have had to adapt to the hot, dry summers and the warm, wet winters (sketch **A**).

The growth of most Mediterranean plants begins with the start of the rainy season in autumn. Bulbs and scrub which have lain **dormant** (inactive) during the summer heat and drought put out new shoots and begin to flower. The seeds of many annual plants germinate. They continue to grow slowly through the winter when water is available and temperatures are warm. They flower in the spring when temperatures get warmer and while the soil is still damp. Seeds ripen in summer. They have thick coats as protection against the intense heat. Green plants, meanwhile, shrivel up to become stiff and thorny (sketch **A**). The evergreen trees grow slowly throughout the year.

How has vegetation been changed by people?

Before people began to live around the Mediterranean Sea the land was covered in forest. Today very little is left and the main vegetation is scrub. This change can partly be blamed on a natural cause – a decline in rainfall. Mostly it is the result of activity by people who settled here, and their grazing animals. The trees were cut down for building houses and ships, and to be used as fuel. Young trees were eaten by herds of sheep and goats. Forest fires, some started deliberately, have added to the destruction. Some pine and cypress trees still grow in remote areas (photo **B**). Where the land is not used for farming or building it is either covered in a thick tangle of low, spiky shrubs or left as bare rock (photo **C**).

B Mediterranean woodland, Crete

C Mediterranean scrub, Crete

Activities

1. Name six Mediterranean plants.
2. **a** Why do Mediterranean plants grow mainly in winter?
   b Why do they only grow very slowly in summer?
3. Give three ways by which the Mediterranean vegetation can survive in the hot, dry summer.
4. Give three ways in which the natural Mediterranean vegetation has been altered by human activity.
5. The olive (diagram **D**) is an important tree and crop in many Mediterranean countries. Make a large copy of diagram **D** and complete it by adding the following labels:

long roots | thick trunk | small leaves | fruit with a thick skin | thin soil

D

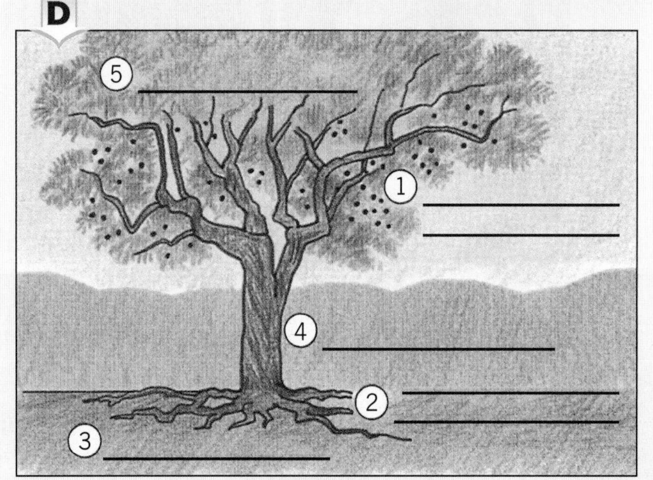

Summary

Mediterranean vegetation has had to adapt to hot, dry summers and warm, wet winters. It has been affected by human activities such as the felling and burning of forests and the keeping of grazing animals.

What is soil erosion?

Soil is one of our most important resources. We depend on it for most of the food that we eat and without it we would be unable to survive. Yet each year some 75 million tonnes of soil are lost around the world. The removal of soil from one place and its deposition elsewhere is called **soil erosion**.

A

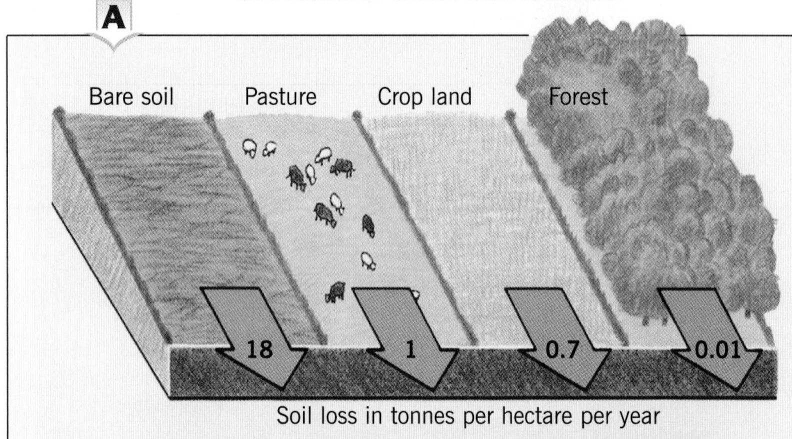

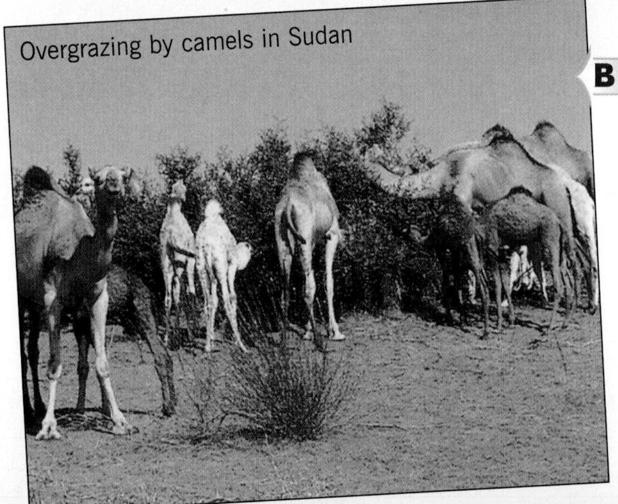
Overgrazing by camels in Sudan

B

C Soil erosion

Soil is usually blown away by the wind or washed away by water running over the ground's surface. Erosion is greatest on steep slopes and when the soil is bare. It is least when there is a thick cover of vegetation. This is because plants and trees provide shelter from rain and wind and their root systems hold together the soil particles, making them difficult to remove. Diagram **A** shows how different types of ground cover are affected by soil erosion in eastern England.

Soil erosion is a natural process but in some places it has been increased by bad farming methods.

- **Overgrazing**
  Sometimes too many animals are kept in one area. They eat all the vegetation and it dies off. This leaves the ground bare and unprotected as in photo **B**. Wind and rain can then carry off the loose soil.
- **Up and down ploughing**
  Farmers find it easier to plough up and down a slope rather than across it. When it rains, water flows straight down the furrows and takes with it large amounts of soil. The furrows can quickly turn into deep gullies like those in photo **C**.
- **Deforestation**
  This is the clearing away of forests, usually so that the land can be used for growing crops. Once the trees have gone there are no leaves to protect the soil from rainfall and no roots to hold the soil in place. This makes it easy for the soil to be washed or blown away.
- **Soil exhaustion**
  Sometimes the soil is overused by the growing of too many crops. Eventually it loses its goodness, crops can no longer grow and the bare soil is quickly removed by the action of wind and water.

D Some effects of soil erosion

Activities

1. Why is soil so important to us?
2. Draw a tree and label it to show how it can help prevent soil erosion.
3. Describe how each of the farming methods shown in diagram **E** can cause soil erosion. You could add drawings for each one to make your descriptions clearer.
4. Copy table **F** below and complete it by writing the effects of soil erosion in the correct columns. Write no more than six words for each effect. Some effects may go in both columns. Give your table a title.

F

During a period of drought and wind	After a day of heavy rain

E

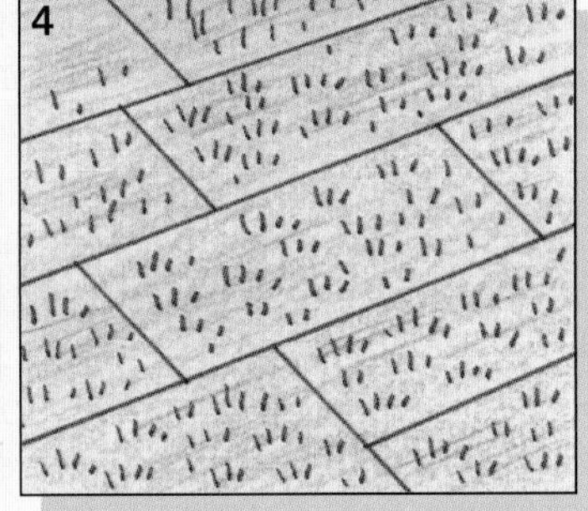

EXTRA

Imagine you are head of the village shown in diagram **D**. Write a letter to the farmer in the area whose methods are causing soil erosion. Explain the effects that his methods are having and suggest how they might be reduced.

Summary

Soil erosion is the removal of soil by wind or water. Where leaves intercept the rainfall and roots bind the soil together, erosion is slow. Where people and animals have removed the vegetation cover, soil erosion can be a serious problem.

Soil erosion in Nepal

For some countries soil erosion is a major **hazard**. It may not be spectacular like volcanic eruptions or earthquakes but it can still cause problems for people and damage to the environment. The mountain kingdom of Nepal is an example of a country that is badly affected by soil erosion.

A

The world's highest mountains, the Himalayas, are here in Nepal.

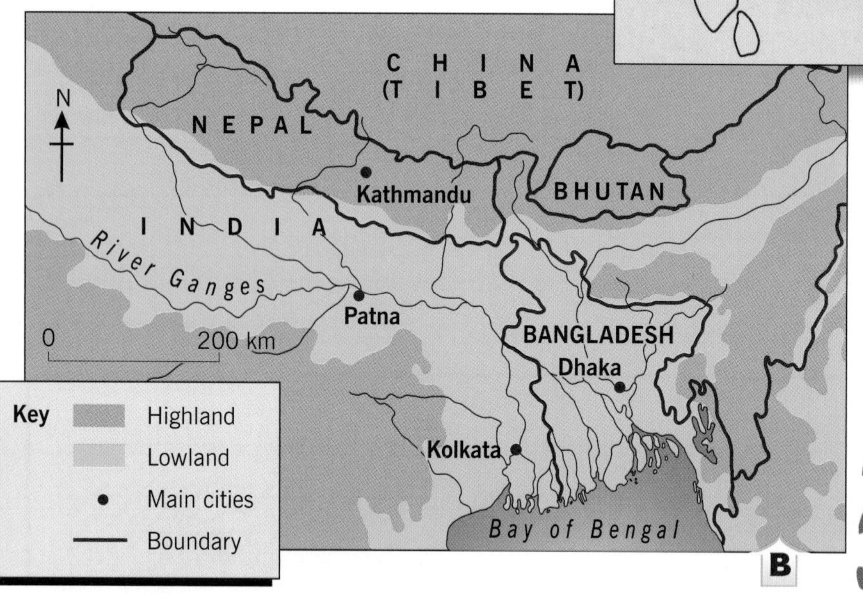

B

Nepal Fact File

Location	Between China (Tibet) to the north and India to the south
Population	25 million
Size	About the same size as England
Occupations	Mainly subsistence farming. Small farm plots on land scattered along the steep-sided valleys
Wealth	One of the world's poorest countries
Physical	One of the most mountainous countries in the world. The climate is very wet in summer and very dry in winter.

Soil erosion has always been a problem in Nepal because of melting snow, heavy rain, unstable soil and steep slopes. Recently, however, the problem has become worse. Nepal's population has doubled in the last 30 years and tourism has increased considerably. This has led to changes in land use and management which in some cases have been damaging to the environment. In the worst cases whole hillsides have been cleared of trees to provide firewood and additional farmland for the extra people. Severe soil erosion has quickly followed as the bare unprotected soil has been washed away by the heavy summer rain.

The increase in erosion has not only affected Nepal but has also caused problems for nearby India and Bangladesh. Due mainly to deforestation the amount of water in the streams and rivers has increased dramatically. This has caused flooding in many places with a loss of farmland, breakdown of communication links and already the death of thousands of people.

C Soil erosion in Nepal

D The erosion problem in Nepal

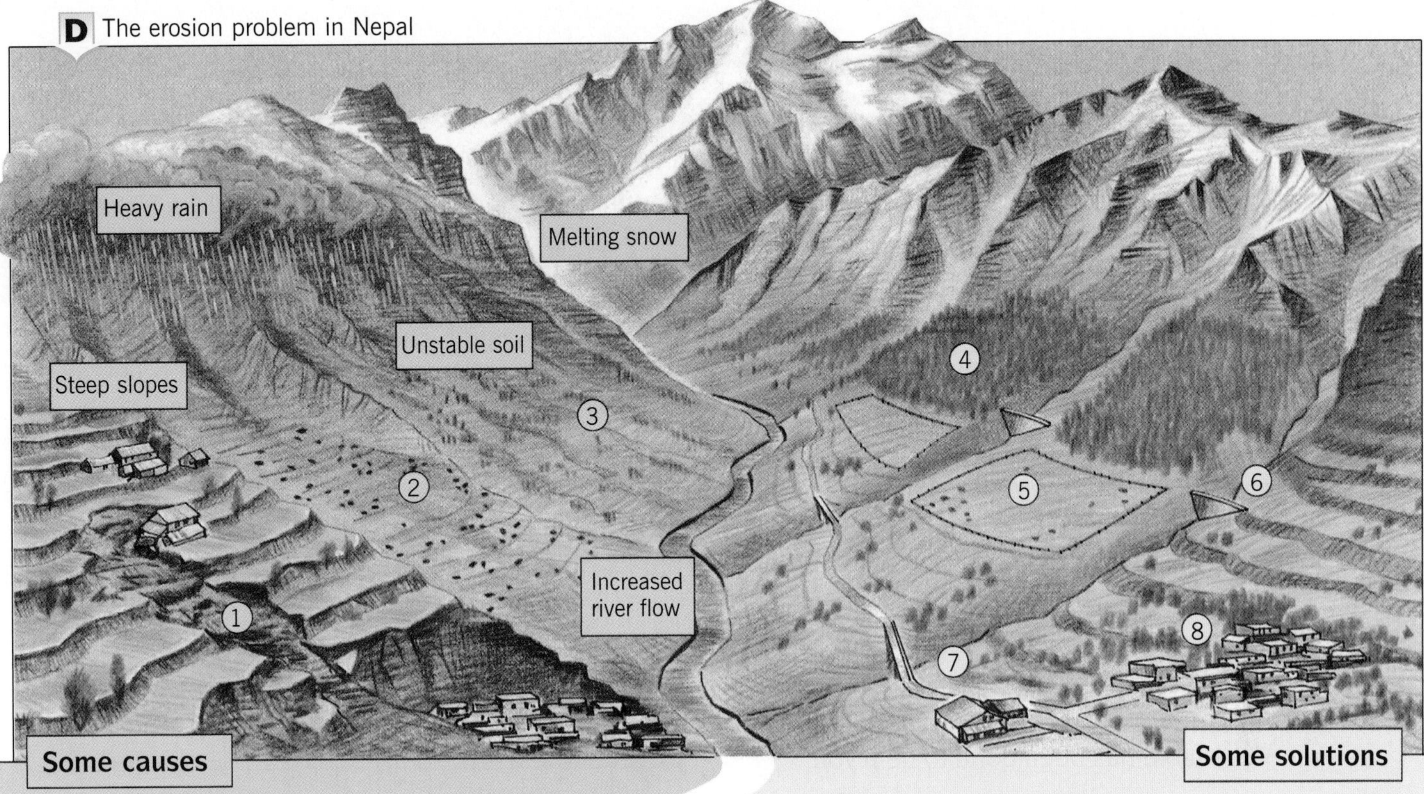

Some causes

① **Collapsed terracing** Farming is only made possible by terracing the valley sides. On steep slopes the terracing may collapse in heavy rain.

② **Overgrazing** In Nepal animals are important for transport, ploughing and food. With the increase in population farmers began keeping too many animals. These have killed off the vegetation, leaving the ground bare.

③ **Deforestation** One third of Nepal's forest has been lost in the last ten years. Most has been cut down for firewood by the increasing population. Tourists (trekkers) each use more firewood per day than local people use in a week.

Some solutions

④ **Tree planting programme**, started by the Nepalese government, to cover bare slopes and reduce run-off

⑤ **Fenced areas** for grazing of fewer animals

⑥ **Small dams** built across gullies to control water flow

⑦ **Government agencies** set up in the valleys to encourage farmers to protect their remaining soil

⑧ **Self-help schemes** for villagers to plant and care for new trees in their own locality

Activities

1. What are the causes of soil erosion in Nepal? List your answers under the headings:
   - Natural processes
   - Human activities.
2. Suggest solutions to any two causes of soil erosion in Nepal brought about by changes in land management.
3. Explain the two newspaper headlines below. Write about 50 words for each one.

Nepal blames tourists for erosion

Bangladesh blames Nepal for floods

EXTRA

- Draw a map to show Nepal, India and Bangladesh.
- Mark the Himalayas and the main rivers.
- Name the cities of Kathmandu, Dhaka, Kolkata and Patna.
- Shade in **brown** the areas where soil erosion will be greatest.
- Shade in **green** where deposition and flooding will be most likely.

Summary

Soil erosion in Nepal is a result of both natural processes and the way people use the land. Careful management of the land can help reduce the problem of erosion.

In a natural environment, climate is the main factor affecting the amount and type of vegetation. Temperature, rainfall, sunlight and wind are all important factors affecting plant growth. As we have seen in this unit, different types of climate produce different types of vegetation. This results in different ecosystems, as the animal life that can be supported also changes.

Look at map **B** below which shows an imaginary continent in the northern hemisphere. On it are marked the boundaries of seven major ecosystems. Your task in this enquiry is to determine where each of the ecosystems described in drawing **G** is most likely to be located. The enquiry assumes that there is no interference from people and that the vegetation cover is natural and unaffected by human activity.

How can climate affect the pattern of natural ecosystems?

1 **a** Make a larger copy of table **A**.
b Complete the first five columns of your table using information from maps **B**, **D**, **E** and **F**. Part of the table has been done to help you.

2 Look carefully at drawing **G**, the 'Guide to natural ecosystems'. Decide which ecosystem matches each of the seven locations in your table. When you have done this, write your decisions in the 'Vegetation type' column of your table.

3 **a** Make a larger copy of map **B**.
b Add a key with boxes for the mountain area and seven different ecosystems. Choose appropriate colours for each box.
c Colour the map to show the pattern of natural ecosystems.

4 Imagine that you are going on a long trek from the north of the continent to the south.
a Make a larger copy of arrow **C** and add simple drawings to show the main types of natural vegetation you would find on your journey.
b Suggest reasons for the differences in vegetation cover along your route.

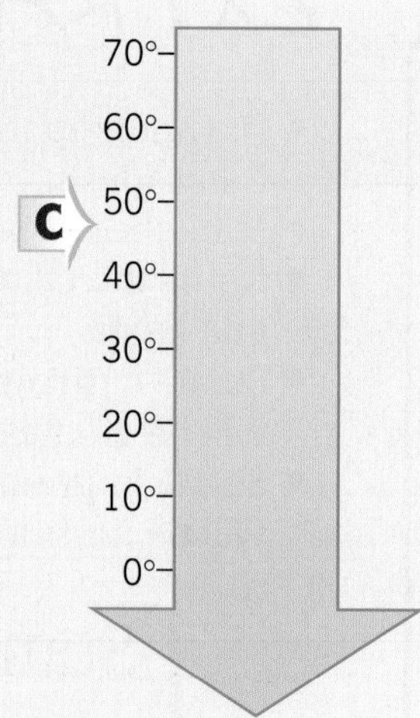

A

Place	Latitude	January temp. (°C)	July temp. (°C)	Annual rainfall (mm)	Vegetation type
A					
B			12°C	700 mm	
C					
D					
E					
F	11°N	22°C			
G					

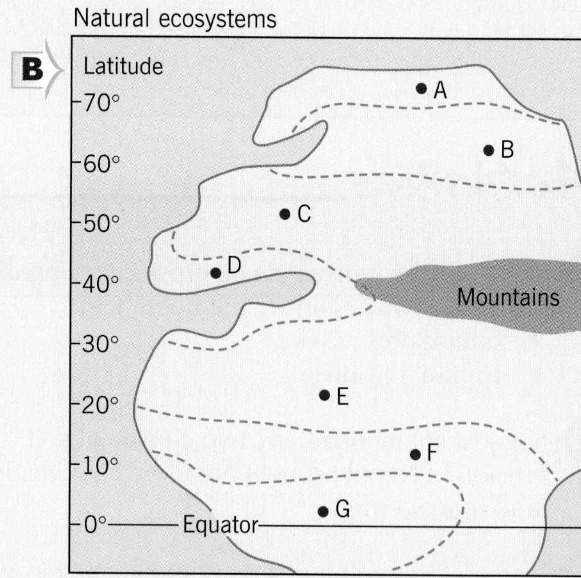

D January temperature (°C)

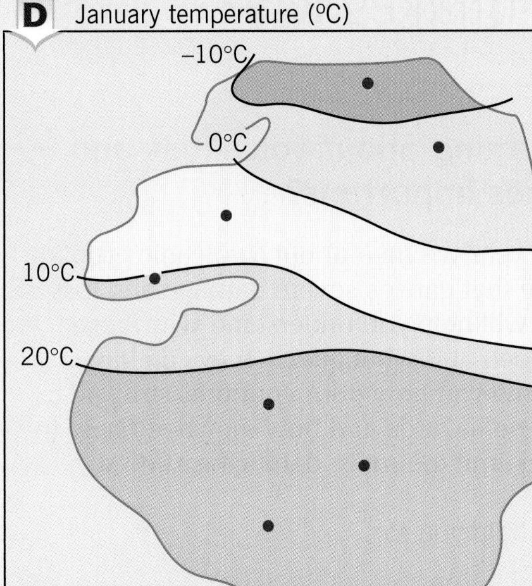

E July temperature (°C)

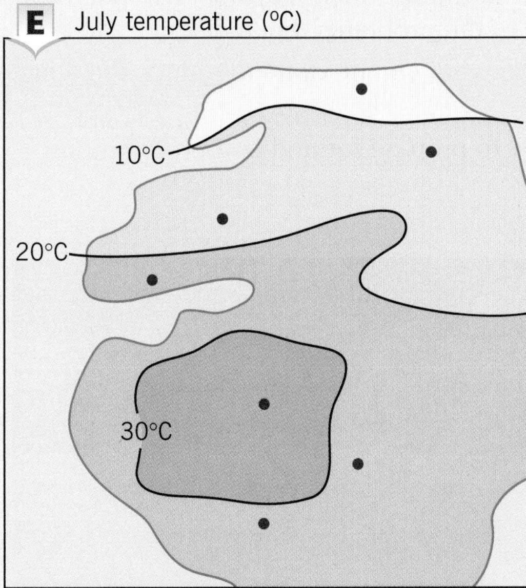

F Annual rainfall (mm)

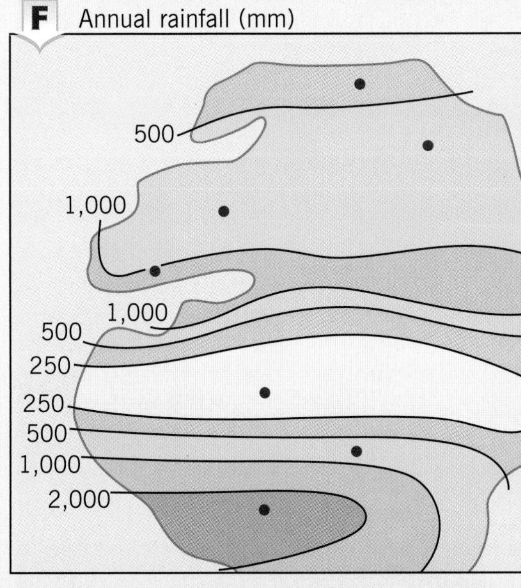

G

Guide to natural ecosystems (biomes)

1 Hot desert
Very high temperatures (up to 45°C) and little or no rainfall (below 500 mm). Mainly cacti, small, thorny bushes and the occasional date palm.

2 Deciduous forest
Cool winters, warm summers and moderate rainfall. Deciduous trees like oak and beech that lose their leaves in winter.

3 Tropical rainforest
High temperatures (up to 30°C) and high rainfall (up to 2,500mm). Dense, luxuriant forest with a huge variety of trees and shrubs.

4 Mediterranean
Hot, dry summers (up to 30°C) and mild, wet winters (10–15°C). Scrub with small, stunted trees and bushes. Some open woodland.

5 Tundra
Very cold and dry. Winter temperatures below –10°C. Mainly mosses, lichens and poor grasses. Some dwarf trees like birch and alder.

6 Grassland
High temperatures (up to 30°C) but little rainfall (500–1,000 mm). Mainly grass with some scattered bushes and trees.

7 Coniferous forest
Cold winters and cool summers. Evergreen trees such as spruce and pine which keep their leaves all year.

2 Volcanoes and earthquakes

What are volcanoes and earthquakes like?

What is this unit about?

This unit explains how movements in the earth's crust can cause both volcanic eruptions and earthquakes. It also looks at how these natural hazards may bring danger to people and cause severe damage to property and the surroundings.

In this unit you will learn about:

- **the distribution of volcanoes and earthquakes**
- **their causes and effects**
- **the different ways in which countries respond to these natural hazards**
- **how the dangers may be reduced.**

Why is learning about volcanoes and earthquakes important?

Almost every year we hear about a volcanic eruption or earthquake that causes severe damage and loss of life. This unit will help you understand why these disasters happen and what effects they can have. It will also show you how poor countries struggle to cope with these hazards and how they need help to save lives and limit the more damaging effects.

This unit will help you to:

- develop an interest in volcanoes and earthquakes
- be aware of the problems that a disaster causes
- understand what can be done to reduce the worst effects
- know how to prepare for and cope with natural hazards.

A Mount Vesuvius and Naples, Italy

B Damage caused by the Kobe earthquake, Japan

C Kilauea, Hawaii

- What is happening in photo **A**? What problems may this cause? What would you do if you and your family lived here?
- What has happened in photo **B**? How will people be affected? What would need to be done immediately after the event and in the months ahead?
- Describe how you would feel if you were the helicopter pilot in photo **C**.

Where do volcanoes and earthquakes happen?

There are thousands of **volcanoes** around the world. Some are extinct, some are dormant and some may be erupting even as you read this book. When they do erupt you can be sure that wherever they are, there will be danger and, probably, damage.

Scientists now know a lot about volcanoes but they still find it difficult to predict exactly where and when an eruption will actually happen. What we do know, however, is that volcanic eruptions do not occur just anywhere on the earth's surface but they are confined to certain areas. Map **A** shows these areas.

Notice that most volcanoes occur in narrow belts or are grouped together in small clumps. One belt runs all the way round the edge of the Pacific Ocean and is called the '**Ring of Fire**'. Another belt runs through the islands of the Indian Ocean. There is also great volcanic activity on Iceland. Can you find the area in the Mediterranean Sea where the volcanoes of Italy are located?

A

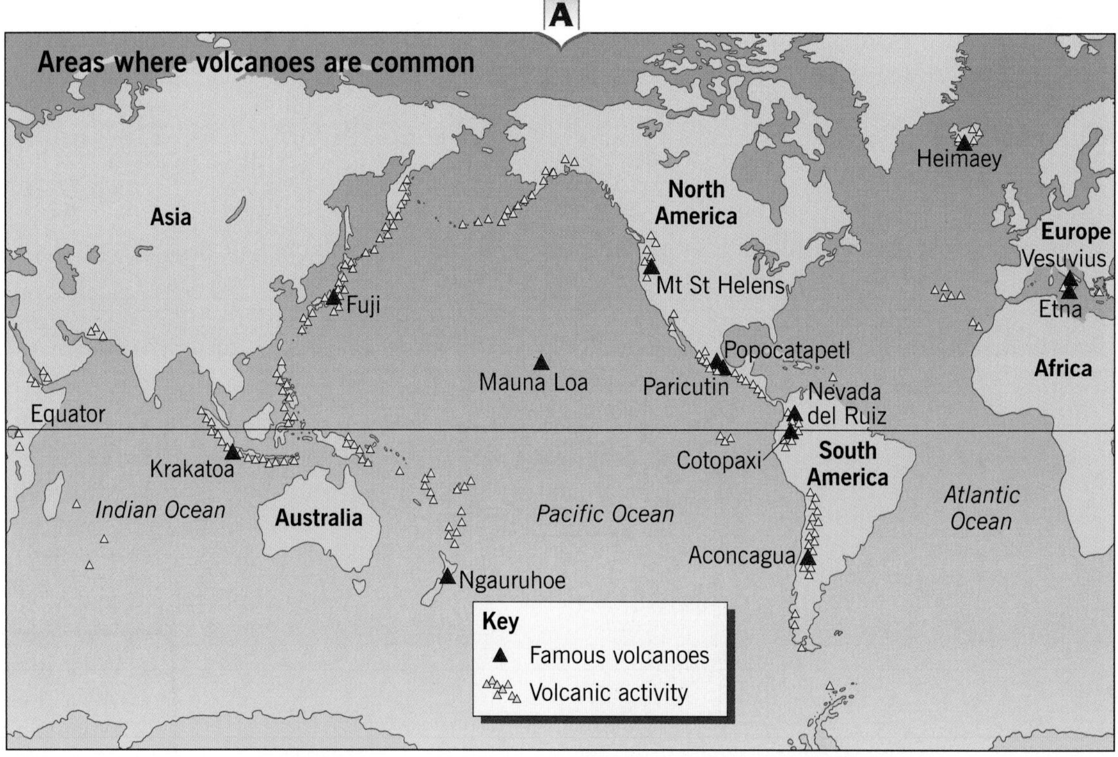

Activities

a Which four of the following describe where volcanoes may be found?

- All over the world
- In narrow belts
- In Iceland
- In Japan
- In Central Asia
- Along the west coast of North and South America
- In Australia

b Name two other places where volcanoes may be found.

Suggest reasons for the name of 'Ring of Fire'.

3 You may need an atlas for this question. Match each of the volcanoes named on map **A** with a country from the list in the box below. Some countries may be used more than once.

- New Zealand • Colombia • Japan
- Mexico • Italy • Argentina
- Indonesia • Iceland • Ecuador • USA

Earthquakes are happening all the time. Some are so weak that they can hardly be felt and instruments called **seismographs** are needed to detect them. Others, like the San Francisco earthquake, are so powerful that the shaking of the ground causes buildings to collapse and landslides to occur. The Indian Ocean earthquake of 2004 caused **tsunami** waves that killed more than 300,000 people.

Earthquakes can occur anywhere, but they are much more common in some places than in others. Map **B** shows where earthquakes regularly happen. Look carefully at their distribution. They are mostly arranged in long narrow belts. One belt goes down the middle of the Atlantic Ocean. Another follows the west coast of North and South America and then goes all the way round the edge of the Pacific Ocean to New Zealand. Try to identify some other belts.

Now compare map **B**, showing earthquakes, with map **A**, showing volcanoes. Notice how similar they are. Look particularly at the 'Ring of Fire' and the Mediterranean countries. From studying maps like these, scientists have concluded that volcanoes and earthquakes often occur in the same places and are usually found in long narrow **zones of activity**. These areas can be the most dangerous places on earth.

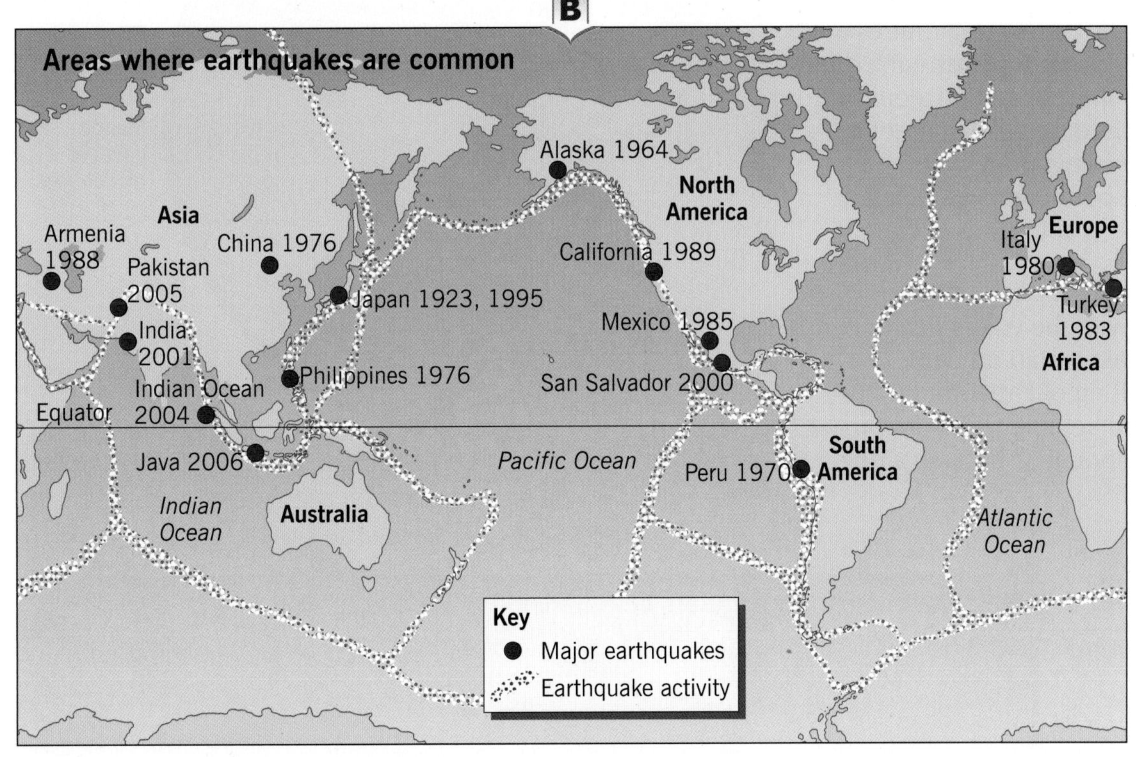

4. Use the information on this page to describe where earthquakes happen.
5. Name five places where a scientist could study both volcanoes **and** earthquakes in the same area.
6. The eastern part of South America is an area largely without volcanoes and earthquakes. Name five other land areas where volcanic eruptions and earthquakes are uncommon.

EXTRA

Use the internet to find out more about one of the volcanoes or earthquakes on map **A** or map **B**. Write a short project about it. Try to include:

- a map to show its location
- a description of the eruption or earthquake
- a list of damage it caused
- labelled drawings to show what happened.

Summary

Most volcanoes and earthquakes are found in long narrow belts across the earth's surface. The main zones of activity lie along the west coast of the Americas and among islands of the Pacific and Indian Oceans.

How do volcanoes and earthquakes happen?

As you have seen on pages 28 and 29, volcanoes and earthquakes often occur in the same places and are usually found in long narrow belts. This gives a clue about how they happen.

The earth was formed 4,600 million years ago. Since then it has been slowly cooling down and a thin **crust** has formed round the outside. The crust is not all one piece but is broken into several enormous sections called **plates**. Some of the plates are as large as continents while others are much smaller. Underneath the crust the rock is so hot that it remains molten and can flow like treacle. The plates float on this layer and move about very, very slowly – just a few millimetres a year. In some places they move towards each other and in others they move apart or scrape alongside each other. The place where two plates meet is called a **plate boundary** (diagram **A**). The movement at these plate boundaries can cause earthquakes and volcanic eruptions to occur.

Look at map **B** which shows the major plate boundaries. Compare it with the volcanoes map on page 28 and the earthquakes map on page 29. Look particularly at the 'Ring of Fire' around the Pacific Ocean. You should be able to see that most of the volcanoes and earthquakes happen along the plate boundaries.

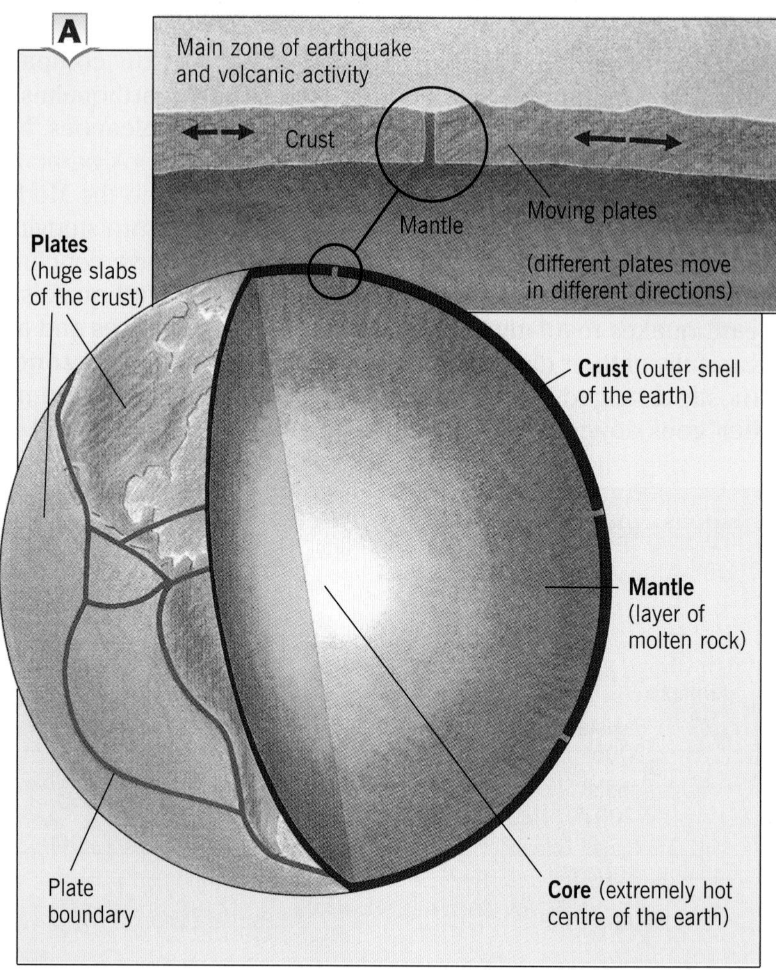

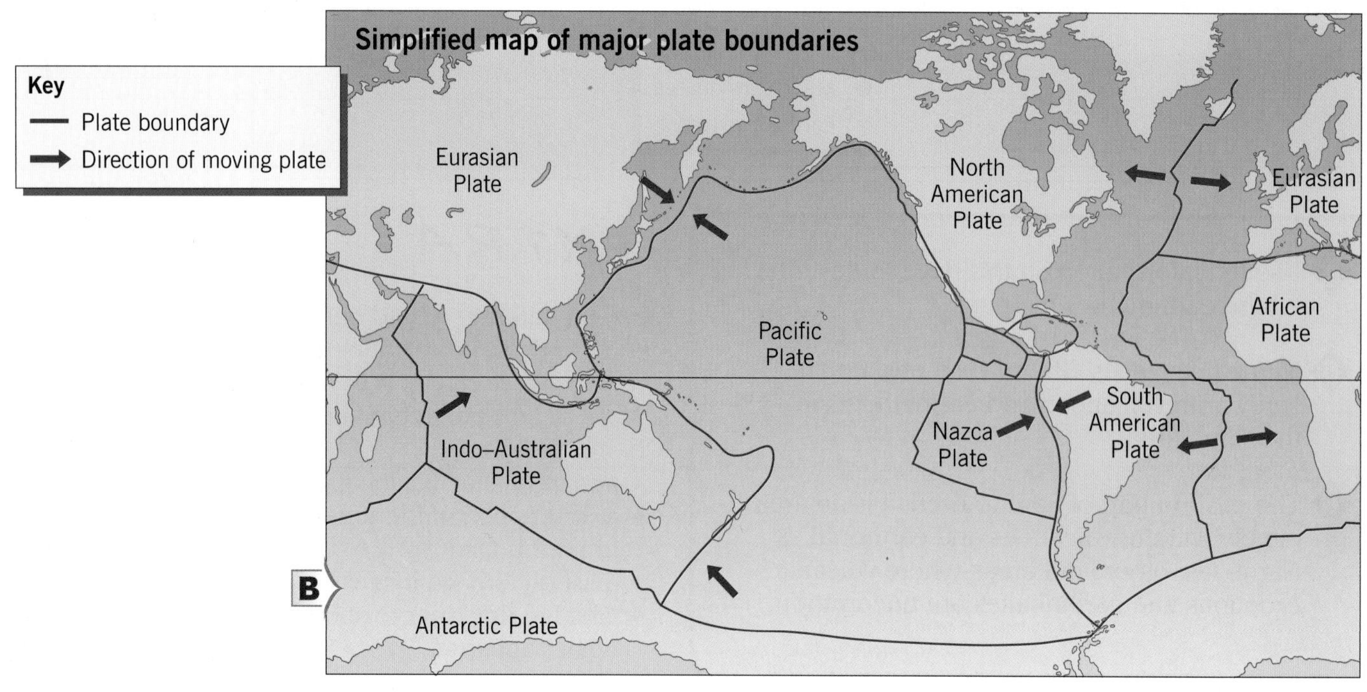

Diagram **C** shows what happens where plates move together. Here, on the west coast of South America, the Nazca Plate is being forced towards and underneath the South American Plate. As the plates move together the friction between them makes the rock melt. The liquid rock (**magma**) rises upwards and erupts on the surface as a volcano. The movement of the plates scraping together also makes the ground shake and sets off earthquakes. South America has over a hundred volcanoes caused in this way and in some places earthquakes happen every day.

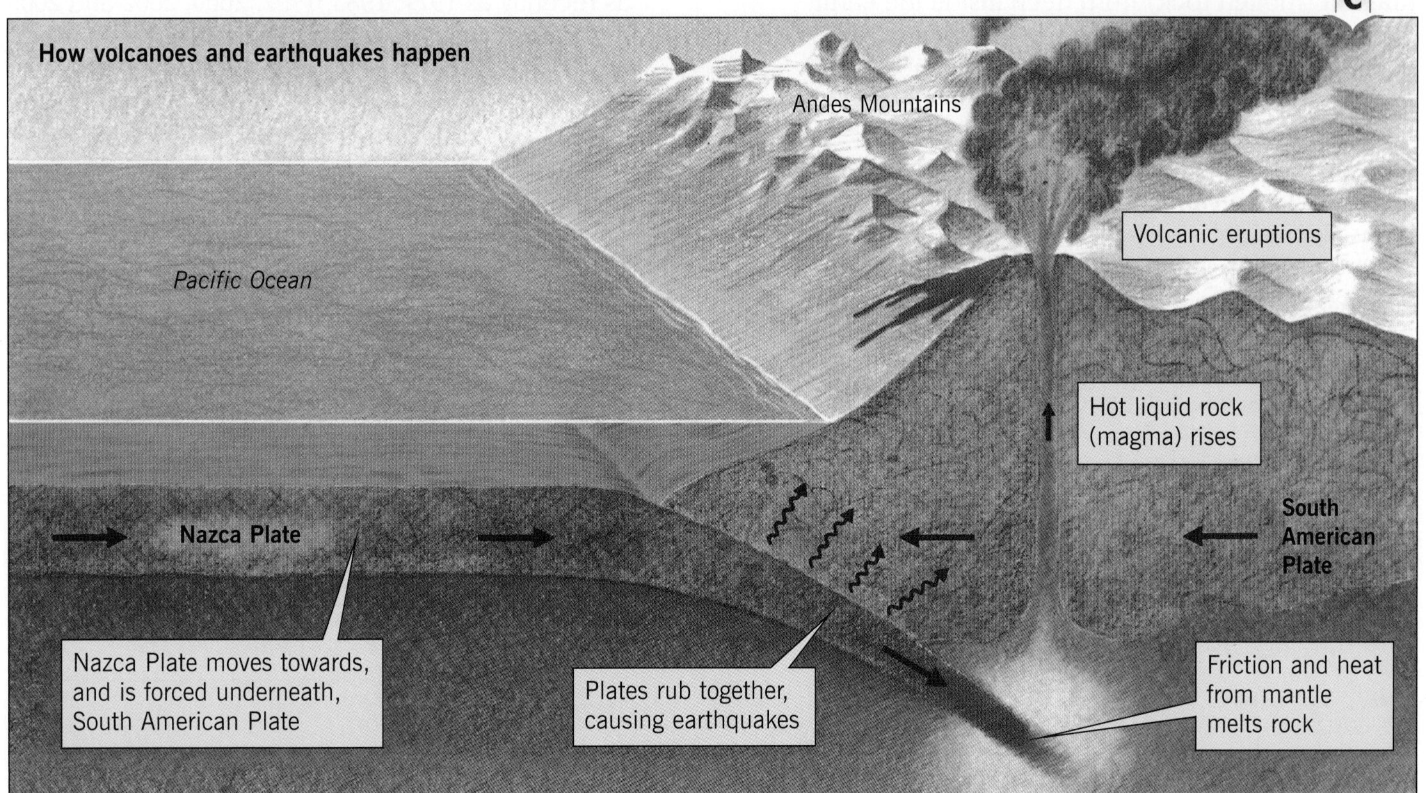

Activities

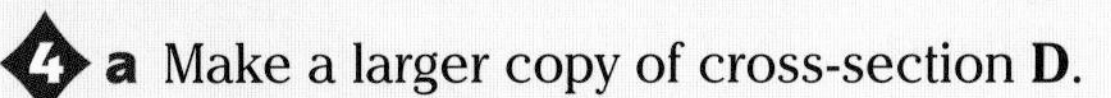

1 Look carefully at map **B**.

a On which plate does Britain lie?

b Why does Britain have no active volcanoes or major earthquakes?

c Which two plates meet along the west coast of the USA?

d Why do earthquakes happen in San Francisco?

2 Sort the statements below into the correct order to show how volcanoes can happen at plate boundaries.

- Molten rock rises
- Friction melts the rock
- Plates rub together
- Volcanoes erupt on the surface
- Plates move towards each other

3 How does the movement of plates cause earthquakes?

4 **a** Make a larger copy of cross-section **D**.

b Name the two plates, the Andes mountains and the Pacific Ocean.

c Draw arrows to show plate movements.

d Put a circle around the zone of activity where there is friction, earthquakes and melting of rock.

e Add a title.

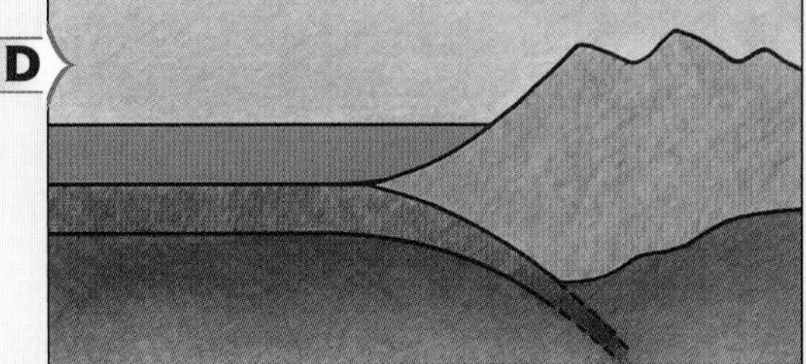

Summary

The earth's surface is made up of several plates that move about very slowly. Volcanoes and earthquakes are most likely to occur in areas where the plates meet.

What are volcanoes?

Volcanoes are openings (**vents**) in the ground where **magma** (molten rock) from deep inside the earth forces its way to the surface. The magma may appear as flows of molten **lava**, as **volcanic bombs**, as fragments of rock or simply as **ash** and **dust**. Mountains that are made of these materials are called volcanoes. Look at diagram **A**. It shows the main features of a volcano and gives an idea of what one looks like inside.

Volcanoes may be **active**, **dormant** or **extinct**.

- If a volcano has erupted recently and is likely to erupt again, it is described as active. There are over 700 active volcanoes around the world. Mount Etna is an active volcano because it erupted as recently as 1971, 1983, 1992, 2000, 2002 and 2005, and is expected to erupt again in the near future.
- Volcanoes that have erupted in the past 2,000 years, but not recently, are said to be dormant or sleeping. These may be dangerous as it is difficult to predict when they are going to erupt again.
- Many volcanoes are unlikely ever to erupt again. They are said to be extinct because they are dead and their volcanic activity is finished. Britain's last volcanoes erupted over 50 million years ago and have mostly been worn away by erosion. The Edinburgh volcano in Scotland, and Snowdon in Wales, are examples of extinct volcanoes in Britain.

A

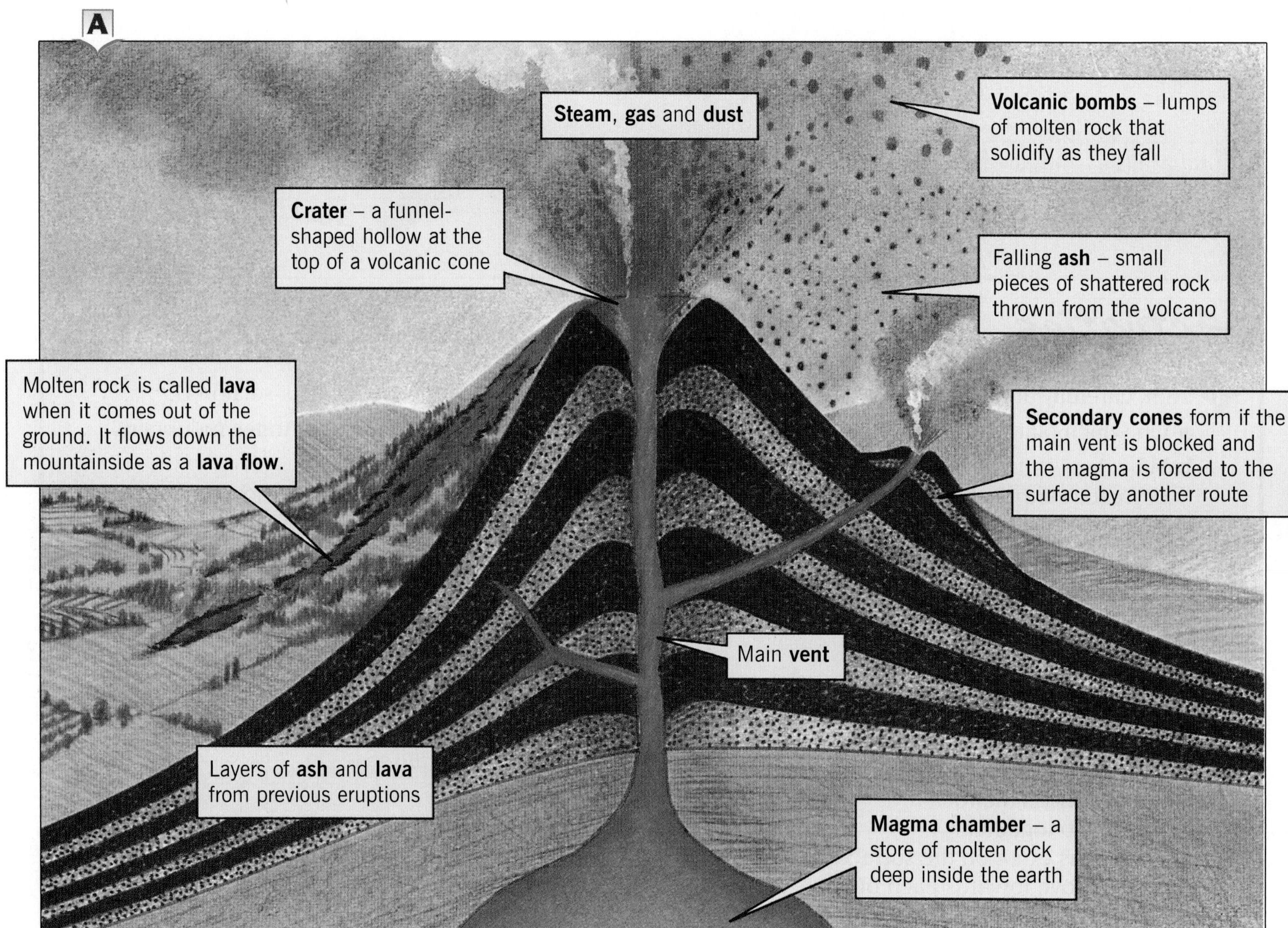

When a volcano erupts, the magma from below the earth's surface rises up the vent to the volcano's crater. It then explodes into the air as ash, dust and volcanic bombs, or flows out as molten lava.
Some eruptions are spectacular and take the form of huge and violent explosions. The greatest volcanic explosion of modern times happened when the Indonesian island of Krakatoa erupted in 1883. The noise from that was so loud that it could be heard over 4,700 km (3,000 miles) away in Australia – see map **A** on page 28.

Not all eruptions are like Krakatoa. Some can be quite gentle and fairly peaceful. Mauna Loa on Hawaii, for example, pours out a steady stream of liquid lava with only a small amount of explosive force and occasional danger to nearby settlements.

Most volcanoes are cone-shaped but the steepness of their slopes can vary considerably. This steepness depends mainly on the type of lava erupted from the vent. Some different types of volcano are shown in diagrams **B**, **C** and **D**.

B

Ash volcano (e.g. Paricutin, Mexico)

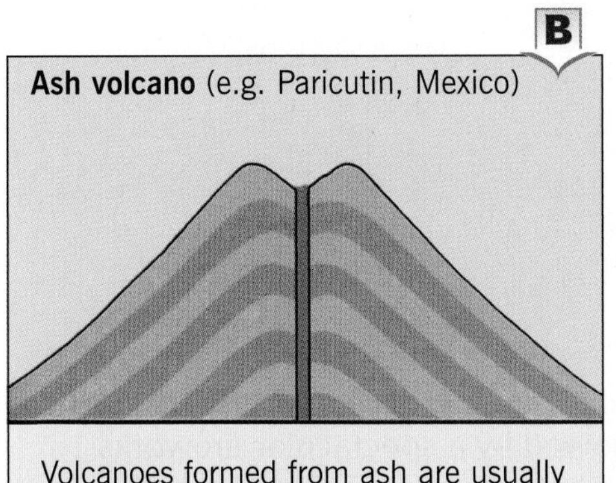

Volcanoes formed from ash are usually steep-sided. They often have a large crater.

C

Lava volcano (e.g. Cotopaxi, Ecuador)

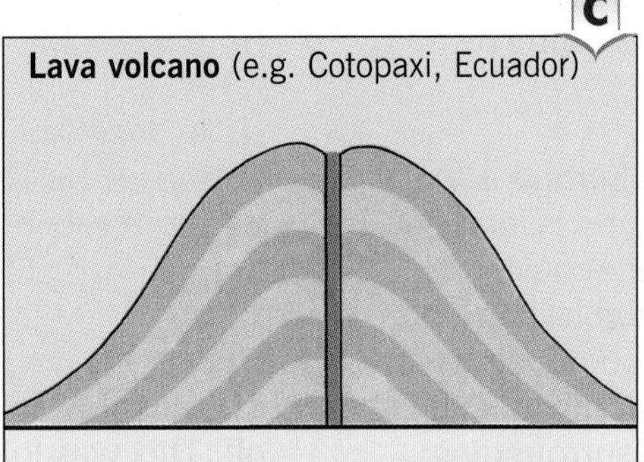

Thick, slow-flowing lava that hardens quickly forms steep-sided volcanoes.

D

Lava volcano (e.g. Mauna Loa, Hawaii)

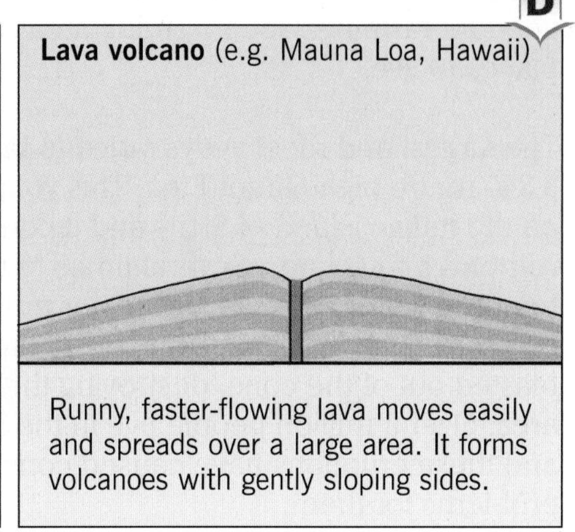

Runny, faster-flowing lava moves easily and spreads over a large area. It forms volcanoes with gently sloping sides.

Activities

1 **a** What are volcanoes?
b What is the difference between active, dormant and extinct volcanoes?

2 Give the meaning of the words below. You could add sketches to some of them to make them clearer and more interesting.

magma | vent | crater | lava | volcanic bombs | volcanic cone

3 Make a simple sketch of photo **E** and label the following:

two craters | steam coming out of vent | volcanic cone | ash and lava

Give your sketch a title.

4 Make a large drawing of a volcano like the one in sketch **A**. Label the features shown in bold type. Underneath your sketch write a description to explain what happens when it erupts. Try to mention all the labelled features.

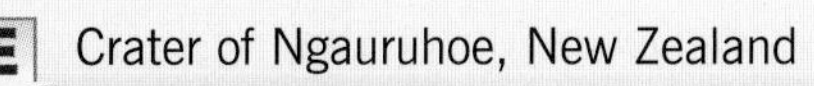

E Crater of Ngauruhoe, New Zealand

Summary

Volcanoes are cone-shaped mountains made from magma that has come from deep below the ground.

What happens when a volcano erupts?

When a volcano erupts it can cause serious problems. People are put in danger and their surroundings can be severely damaged. Problems like this are called **natural hazards**. Earthquakes, floods, drought and strong winds are also examples of natural hazards. In spite of the danger and possibility of great destruction, a lot of people often live in volcanic areas. This is because ash and lava turn into rich fertile soil which is good for farming. Good farming areas are attractive places to live.

A

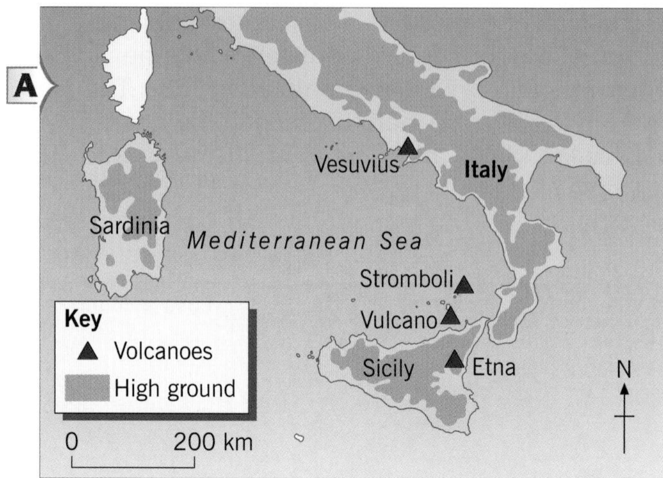

The largest and most active volcano in Europe is the 3,340 metre high Mount Etna. This volcano is located on the Italian island of Sicily and it continuously rumbles and steams. Several times in the last one hundred years Etna has had major eruptions, when masses of ash, volcanic bombs and lava have been blasted out of the cone, destroying the surrounding area. Over a million people live in the Mount Etna area and these eruptions have caused considerable problems for them.

Mount Etna erupted most recently in 1971, 1983, 1992, 2000, 2002 and 2005. In 1971 the eruption began with a huge explosion that sounded like a jet aircraft taking off. This was followed by a spectacular fireworks display when red-hot ash was thrown hundreds of metres into the air and molten lava poured down the mountainside. In this eruption most of the ski slopes and cable car stations were destroyed and a research observatory near the summit was completely wiped out.

B The 'fireworks display' eruption of Mount Etna

The 1983 eruption began in March and continued for several months. Millions of tonnes of lava gushed out of the crater and engulfed a hotel, three restaurants, 25 houses and numerous orange groves and vineyards. The lava flowed at an average speed of 15 km per hour (about the speed you ride a bicycle) and at one time threatened to bury several small villages in its path. Eventually a diversion was made and after a series of controlled explosions, the lava was diverted and the villages saved.

C Damage caused by the lava flows of the 1983 eruption of Mount Etna

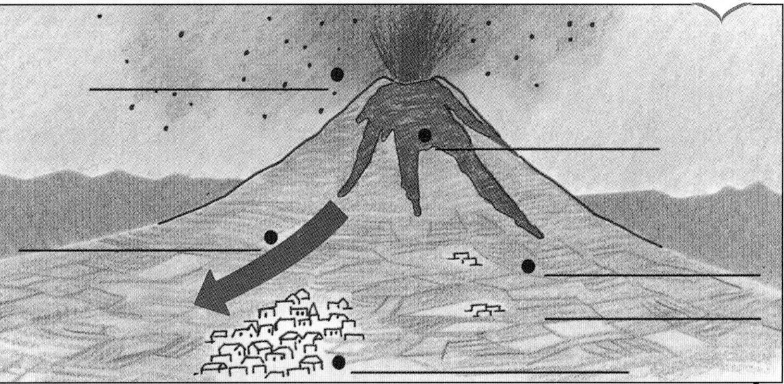

D

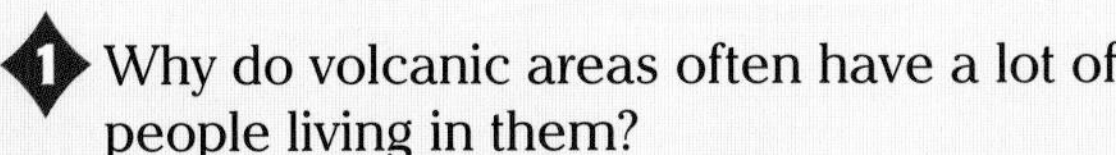

Activities

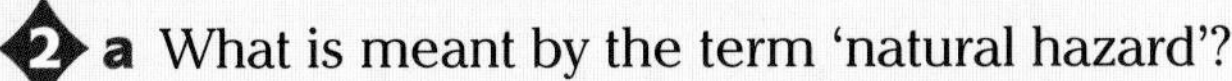

1 Why do volcanic areas often have a lot of people living in them?

2 **a** What is meant by the term 'natural hazard'?

b Give four examples of natural hazards.

3 Describe what happens when Mount Etna erupts by sorting the boxes below into the correct order. Link your boxes with arrows and add a title. You might like to make a simple drawing for each box to make your description clearer and more interesting.

Lava pours down the mountainside	Buildings and property damaged
Rescue service goes into action	Volcano gently rumbles and steams
Loud explosion as volcano erupts	Ash, bombs and lava blasted out of volcano

4 **a** Make a simple copy of sketch **D** and put the following labels in the correct places.

lava flows | threatened settlement | diversion channel | ash and bombs | vineyards and orange groves

b Underneath your sketch list eight problems caused by Mount Etna erupting.

Summary

When volcanoes like Mount Etna erupt they may bring danger to people and cause severe damage to property and the surroundings. Disasters caused by great forces such as volcanic eruptions, earthquakes, floods and strong winds are called natural hazards.

What happens in an earthquake?

On Tuesday 17 October 1989 an **earthquake** hit the Californian city of San Francisco. These two pages explain through photos and newspaper articles what happened on that day.

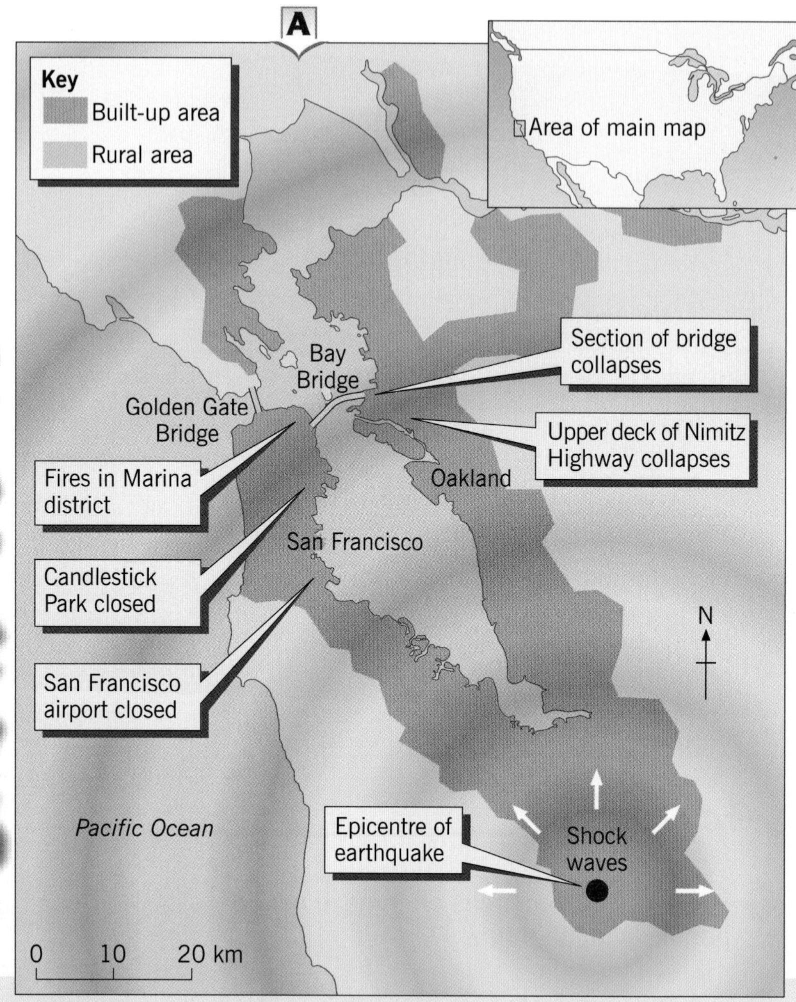

B

Quake hits 'Frisco'

At least 63 people were killed and over 3,000 were injured when an earthquake struck San Francisco at 5.04 p.m. yesterday. It ripped 10ft cracks in roads and a packed highway collapsed, crushing 253 motorists. Rescue workers struggled to free people from damaged buildings while fires roared throughout the city. More than a million homes were plunged into darkness and over 13,000 people were made homeless. Eyewitnesses said they heard a low rumbling noise before the quake hit. Everything then began to shake and buildings started to fall apart. Estimates of the damage already stand at $7 billion. President Bush has promised immediate aid.

Wednesday 18 October, *San Francisco Herald*

C

'The whole world was shaking'

At San Francisco's City Hall, terrified staff dived under desks as pieces of the ceiling and walls came down. Across the street in the health department a giant gusher of water shot through the second floor.

Near the tourist area of Fisherman's Wharf an entire four-storey apartment block collapsed. A woman from the third floor said, 'It seemed like the whole world was shaking. I ran to the stairs to escape but they had gone. There was just a hole. I had to use the fire escape.' Others were not so lucky. They were trapped and couldn't get out.

Fires were inevitable. In one of the worst, a broken gas main exploded turning an entire block in the Marina district into a raging inferno. Firemen were unable to control the blaze which roared on through the night.

Thursday 19 October, *Daily Mail*

Big shake wrecks road

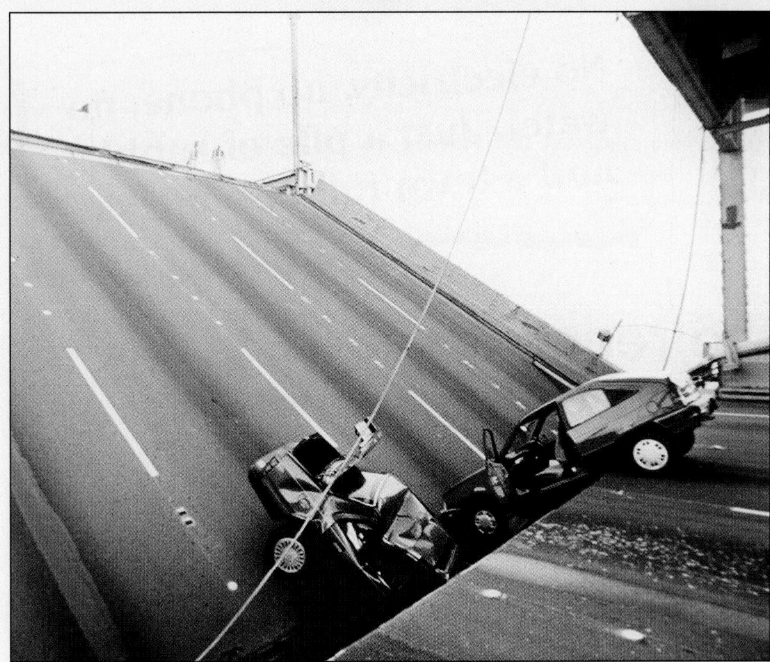

An estimated 200 people were killed when a mile of two-tier road known as the Nimitz Highway collapsed. The road was built to be earthquake proof but the 'big shake' was too much for it. A huge section collapsed onto the road below, squashing hundreds of cars into a space that in places was just 12 inches high.

Further along the highway on the Bay Bridge, more motorists were killed when part of the bridge collapsed. A woman driver on the bridge at the time said, 'The whole structure wobbled and a great gap appeared in the roadway. Cars skidded out of control and some toppled over the edge.'

At Candlestick Park 60,000 baseball fans were packed into the stadium for the game between the Giants and the Athletics. The game had just begun when the earth began to shake and violent tremors ran through the stadium. Cracks opened up in the concrete stands and ripples over a foot high ran right across the park.

Several people were hit by chunks of falling metal and concrete but no-one was killed.

Thursday 19 October, *The Californian*

Activities

1. Complete an earthquake Fact File using these headings.

E

Fact File

Place ____________

Date ____________ Injured ____________

Time ____________ Homeless ____________

Dead ____________ Damage cost ____________

2. Look at the headlines below about the San Francisco earthquake. Write a newsflash to be read out on television giving news about the earthquake. Write about 40 words on each of any four of the headlines given below.

F

Baseball blitzed

Many trapped

Buildings toppled

Fires rage

Bridge collapses

Highway squashed

EXTRA

Below is a list of problems that faced the authorities after the earthquake. Which four do you consider to be the most urgent? Give reasons for your choice.

- Provide new homes for people
- Search for missing people
- Supply medicine
- Rescue stranded people
- Evacuate people in danger
- Supply food
- Supply drinking water

Summary

Earthquakes make the ground shake and may cause buildings and other structures to collapse. Some earthquakes are violent and may cause severe hazards for people.

What happened in the Indian earthquake?

Millions panic as huge earthquake rocks India

No electricity, no phone, no water. Just a pile of rubble and a town full of bodies...

Death toll expected to reach 30,000

At 8.50 am on Friday 26 January 2001, an earthquake struck the heavily populated state of Gujarat in western India. The 'quake measured 7.9 on the Richter scale and sent tremors thousands of kilometres across the continent. They were felt as far away as Nepal and Bangladesh. It was the most powerful earthquake to hit India since 1950 when 1,538 people were killed in Assam.

The 'quake lasted just 45 seconds but caused massive damage. The city of Bhuj, near the epicentre, was worst hit. More than half of the buildings collapsed and most of the town was reduced to piles of rubble. A survivor at the town's Prince Hotel said, 'The whole building began swaying from side to side and everyone started screaming. We ran into the street and watched the building crash down around us'. Elsewhere, whole towns and villages were flattened and electricity and water supplies cut off. Many roads were damaged and closed to traffic.

Most deaths and injuries were caused by building collapse. Rescuers used their bare hands to dig for relatives trapped in the rubble, usually with little success. Many children were killed when at school, whilst a group of 350, taking part in a Republic Day march, were buried when a wall collapsed. Over a million people have been made homeless and are without food or shelter. They face disease and starvation unless international help can save them.

Adapted from a national newspaper article, Monday 29 January 2001

B Rescuers dig out survivors with their bare hands

C Location of earthquake

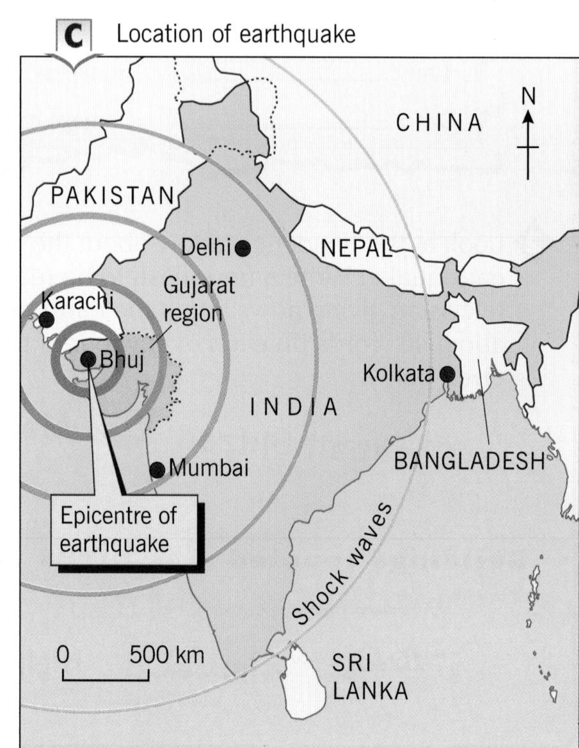

D Earthquake damage

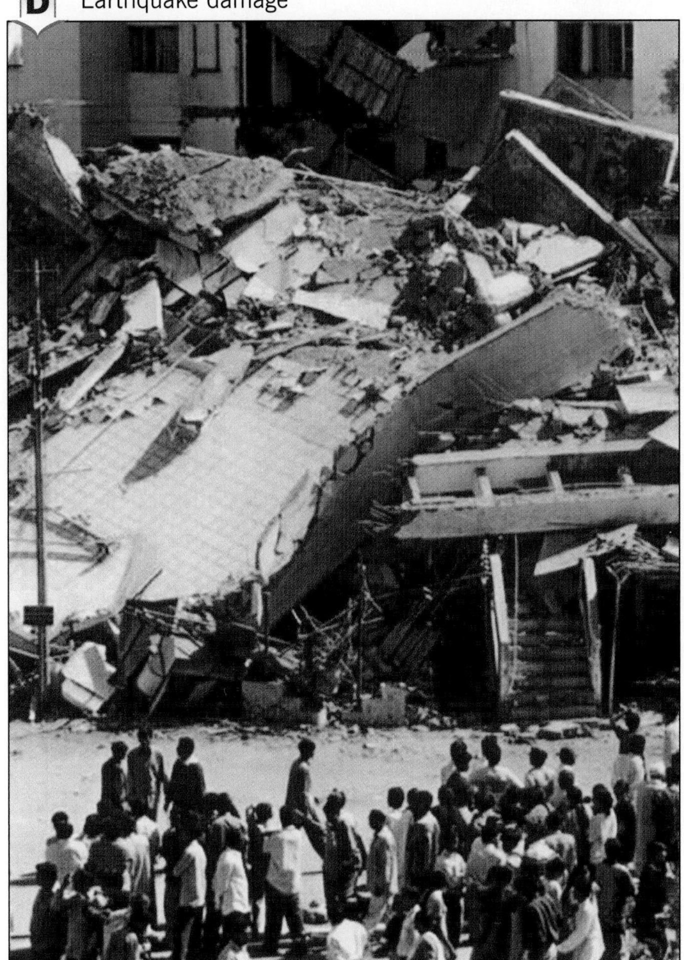

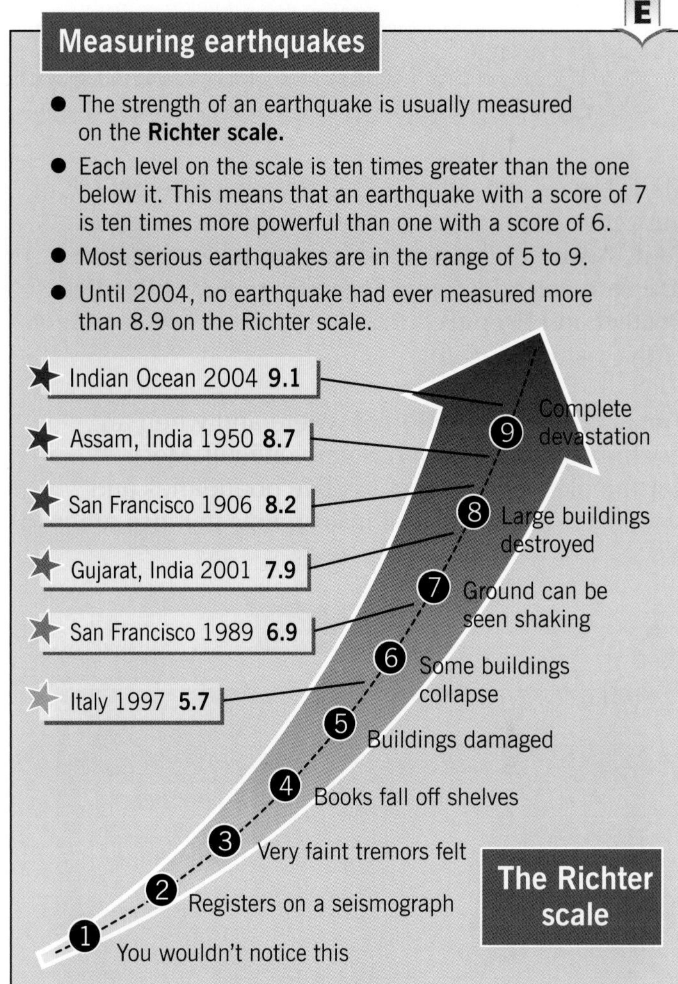

Activities

1 **a** What is the Richter scale?

b Approximately how many times more powerful than the 1989 San Francisco earthquake was:
- the Gujarat earthquake
- the Assam earthquake?

2 Write a newspaper report on the Gujarat earthquake using the headings below.

F

Gujarat earthquake report

a What happened?
b Where did it happen?
c When did it happen?
d What damage was done?
e What were the effects on the people of the area?

3 Shahida Ali was trapped in the ruins of her school for 36 hours before rescue. Describe her ordeal. Try to include the words in the box.

shaking • crashing • screaming • darkness • silence • pain • frightened • crying • rescuers digging • safety • cheering

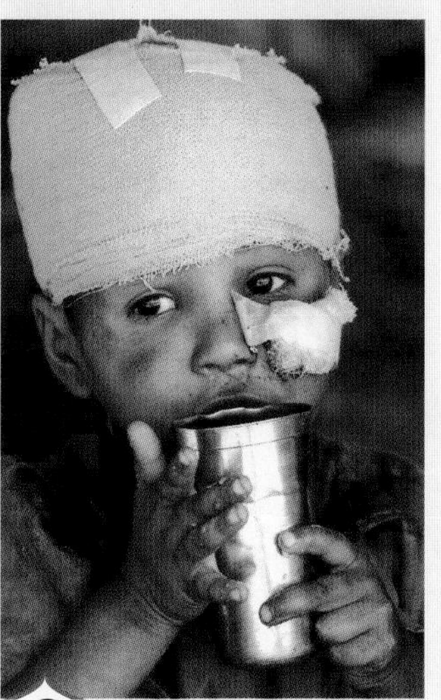

G Shahida Ali survived the earthquake

Summary

Earthquakes are a major natural hazard and usually occur without warning. They can cause considerable damage and loss of life.

How can the earthquake danger be reduced?

Many towns and cities in California have been built in the active earthquake zone along the San Andreas Fault. Aware of the dangers caused by earthquakes, the state has adopted a 'Three Ps' policy of **Predict**, **Protect** and **Prepare**, to try to reduce the worst effects of this natural hazard.

The accurate **prediction** of where and when an earthquake may happen is very difficult. Most earthquakes occur close to plate boundaries and scientists set up sensitive instruments in these areas to monitor changes in the earth. Figure **A** lists some of the warning signs that help the scientists forecast where and when an earthquake might strike.

Most loss of life and damage to property in an earthquake is due to the collapse of buildings. The second part of the 'Three Ps' policy is to design and build structures that are safe and provide **protection** rather than cause danger in an earthquake. San Francisco's TransAmerican Pyramid, shown below, is an example of an earthquake-proof building.

A Predict

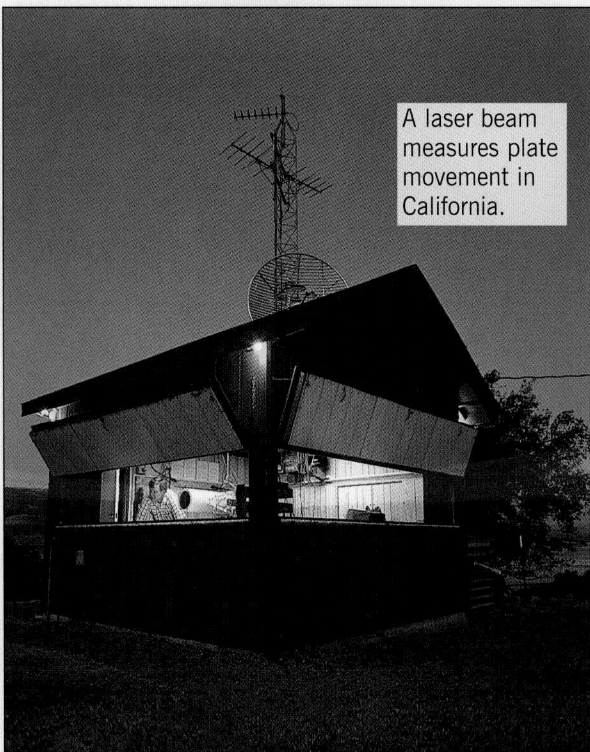

1 Earthquakes are most likely after long periods without any plate movement.
2 Just before a 'quake, small cracks develop in the rock.
   - The cracks cause the rock to swell and bulge.
   - Radon gas seeps out and can be measured as it bubbles to the surface.
   - The cracks fill with water and cause nearby water levels to change.
3 There will be many small foreshocks before the main 'quake. These can be measured with a **seismograph**.
4 Animals often act strangely. Snakes and rats crawl out of their holes and dogs howl.

B Protect

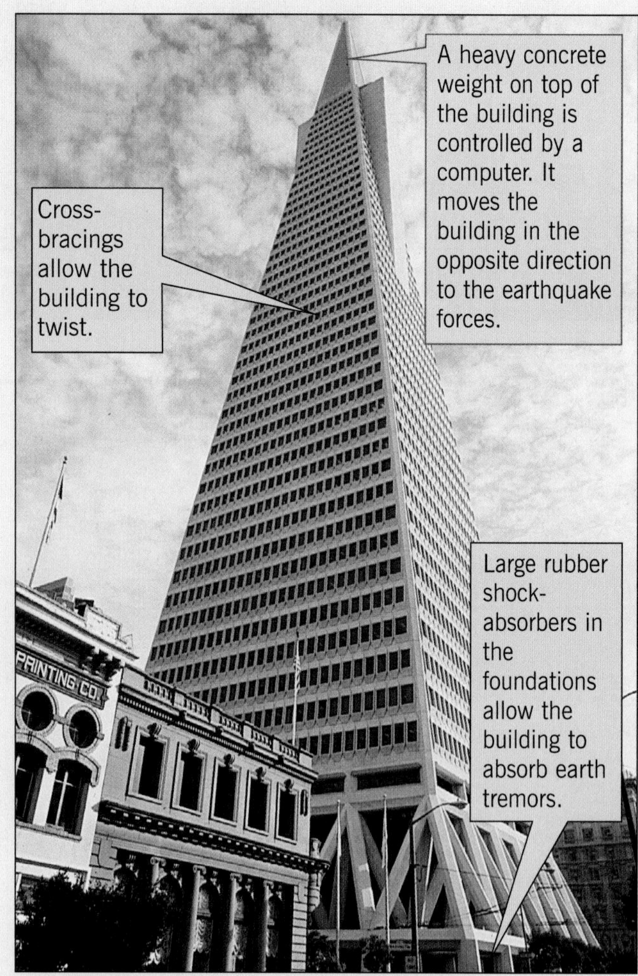

1 All new buildings must comply with strict earthquake planning regulations.
2 Building regulations must be adhered to and frequent safety checks carried out.
3 Existing buildings, roads and bridges should be strengthened.

Good **preparation** and planning can help limit the worst effects of an earthquake. This should involve local authorities and emergency services as well as people living in the area.

Most places that are in danger areas have an emergency disaster plan which is usually in three parts. The first prepares the area for the disaster. The second tries to save lives and look after the people worst affected. The third aims to bring the area back to normal as quickly as possible.

In some parts of California, schoolchildren practise earthquake drills as part of their lessons. At the sound of a bell everyone must shelter under desks and then move quickly outside to be counted.

C

Prepare

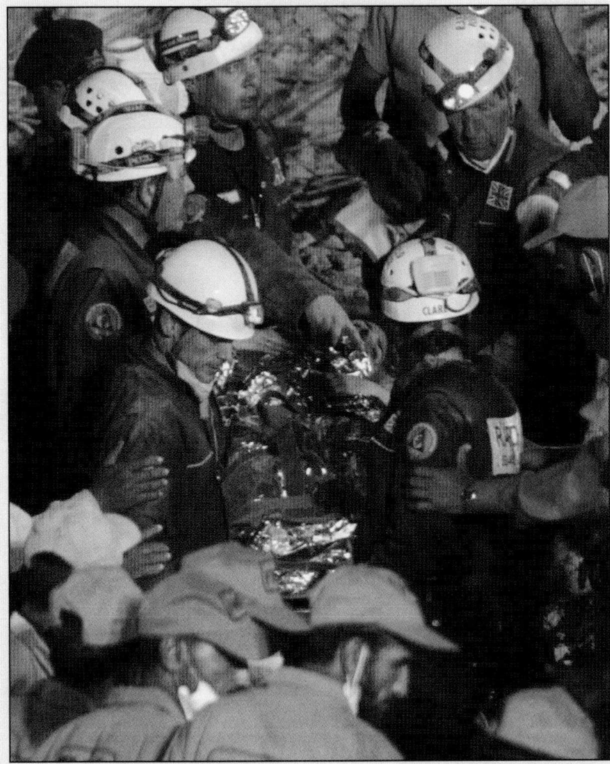

1 Prepare disaster plans and carry out regular practices.
2 Train emergency services such as police, fire and ambulance crews.
3 Organise and prepare hospitals and evacuation centres in safe areas.
4 Educate people on what to expect and what will happen – turning off the gas supply, for example.
5 Organise emergency supplies of water, food and power in advance.
6 Set up an efficient earthquake warning and information system.

Activities

1 It is not easy to predict an earthquake.

a What does 'predict' mean?

b Draw a star diagram to show four signs that suggest an earthquake may be about to happen.

2 Write out the sentence beginnings in drawing **D** and complete them with the correct endings from the following list:
... follow the rules for safe buildings.
... practise what to do in an earthquake.
... include food, clothing, a radio and torch.
... not be built in earthquake zones.
... be prepared to give out earthquake advice.
... help people who get injured.

D

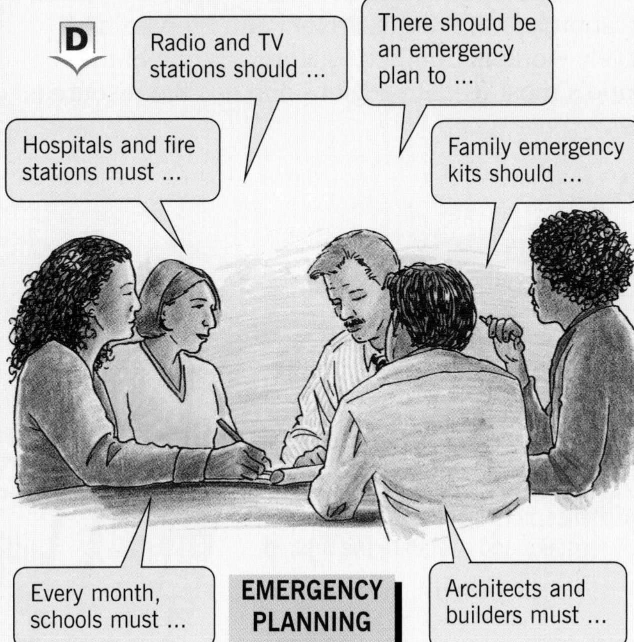

3 **a** List the different ways that could be used to inform people about earthquakes.

b Draw a poster for your classroom wall to show exactly what should be done in an earthquake. Add drawings to make it clearer and more interesting.

Summary

It is impossible to prevent earthquakes from happening. A policy of prediction, protection and preparation can help save lives and reduce damage to property.

Two earthquakes compared

Despite the best preparations and well-practised emergency plans, earthquakes still kill people. This is mainly because it is impossible to predict exactly where or when an earthquake will strike, and how powerful it will be. For example, scientists are sure that certain areas of California will have more earthquakes, but despite all their efforts they still cannot say when.

However, as we have seen on pages 40 and 41, there are many other ways of limiting the worst effects of earthquakes. The San Francisco 'quake of 1989 was very powerful yet there were few deaths and the only buildings to collapse were those that had been built on unstable, reclaimed land. San Francisco's 'Three Ps' approach of Predict, Protect and Prepare had largely worked. But then California is one of the world's most wealthy regions and has the resources and huge sums of money that a successful disaster plan requires.

Sadly, many countries in earthquake zones are not so fortunate and are just too poor to protect themselves against natural disasters such as this. Some have no disaster plans at all whilst others are unable to enforce building safety codes or provide adequate resources and training for their emergency services.

In the Indian earthquake of 2001, for example, thousands of people died as poorly constructed buildings collapsed and emergency services were unable to cope with the disaster. Limited medical facilities and poor transport worsened the problem. The Indian government was seriously short of money, and had to ask for £1 billion of international aid to support the rescue effort.

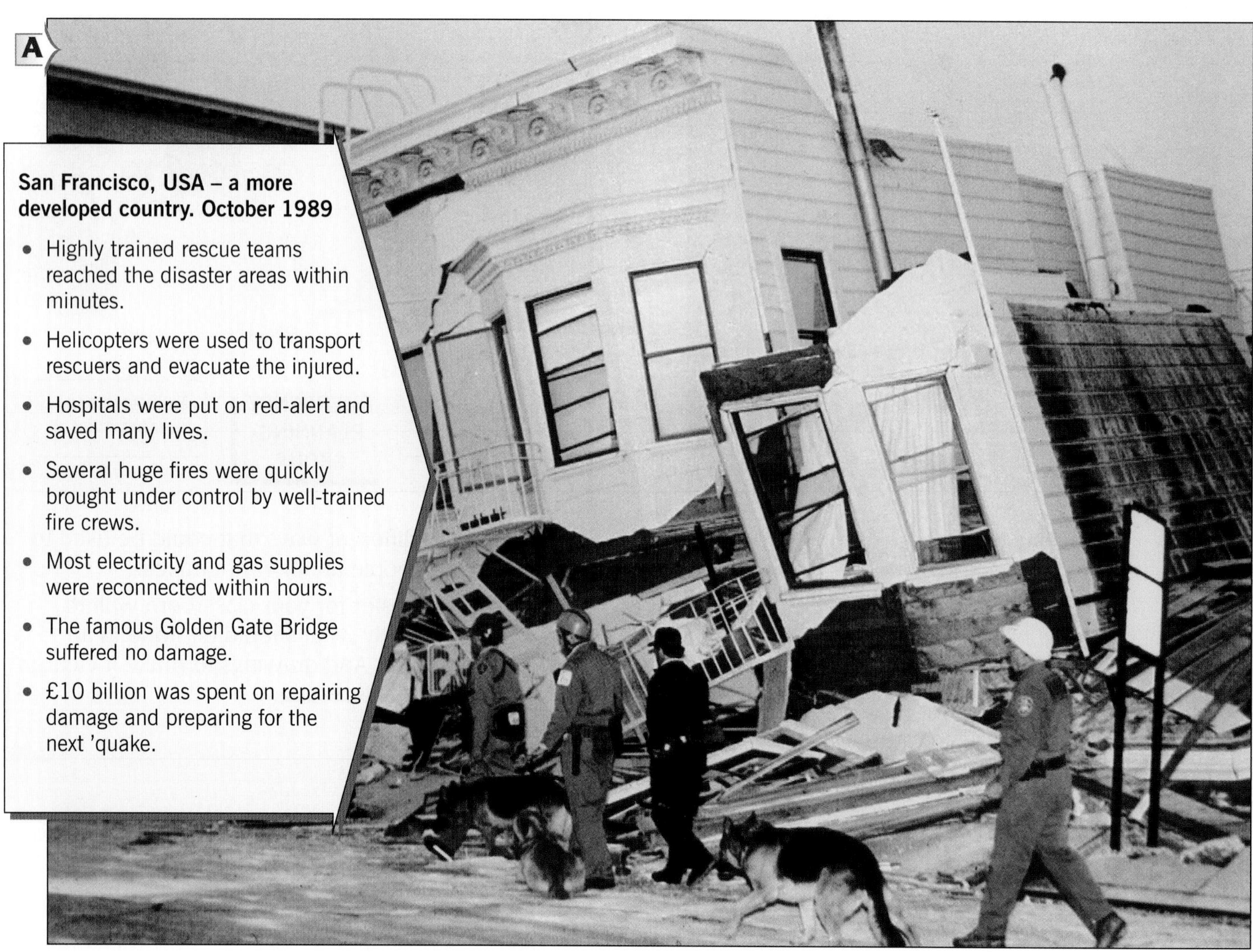

A

San Francisco, USA – a more developed country. October 1989

- Highly trained rescue teams reached the disaster areas within minutes.
- Helicopters were used to transport rescuers and evacuate the injured.
- Hospitals were put on red-alert and saved many lives.
- Several huge fires were quickly brought under control by well-trained fire crews.
- Most electricity and gas supplies were reconnected within hours.
- The famous Golden Gate Bridge suffered no damage.
- £10 billion was spent on repairing damage and preparing for the next 'quake.

Activities

1 Why is it not possible to prevent all damage and loss of life in an earthquake?

2 **a** Make a larger copy of table **B**.

b Assess the success of each country's earthquake preparations by writing *Good* or *Bad* in the first column.

c Give a reason for your choice in the second column.

B

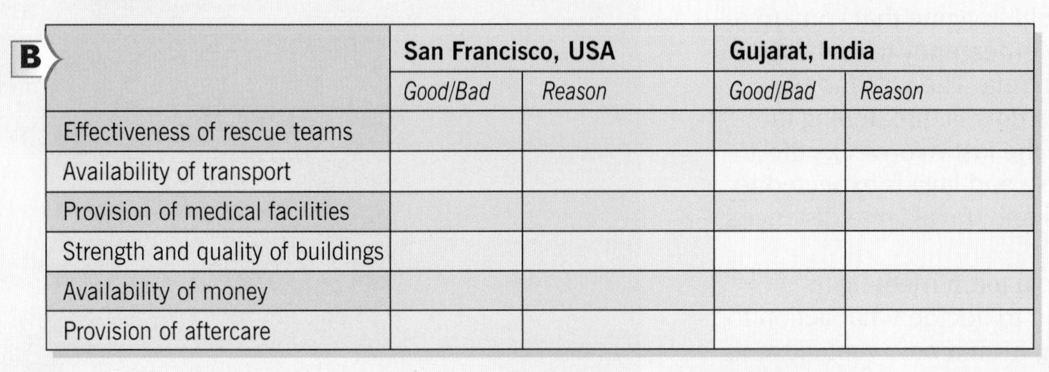

	San Francisco, USA		Gujarat, India	
	Good/Bad	*Reason*	*Good/Bad*	*Reason*
Effectiveness of rescue teams				
Availability of transport				
Provision of medical facilities				
Strength and quality of buildings				
Availability of money				
Provision of aftercare				

3 Explain why earthquakes in poorer, **less developed** countries usually do more damage than similar earthquakes in richer, **more developed** countries.
Use these headings:

- Prediction
- Protection
- Preparation.

Summary

Poor countries, like India, find it very difficult to cope with natural hazards such as earthquakes. The effects of these disasters are therefore a lot worse than they would be for a rich country.

C

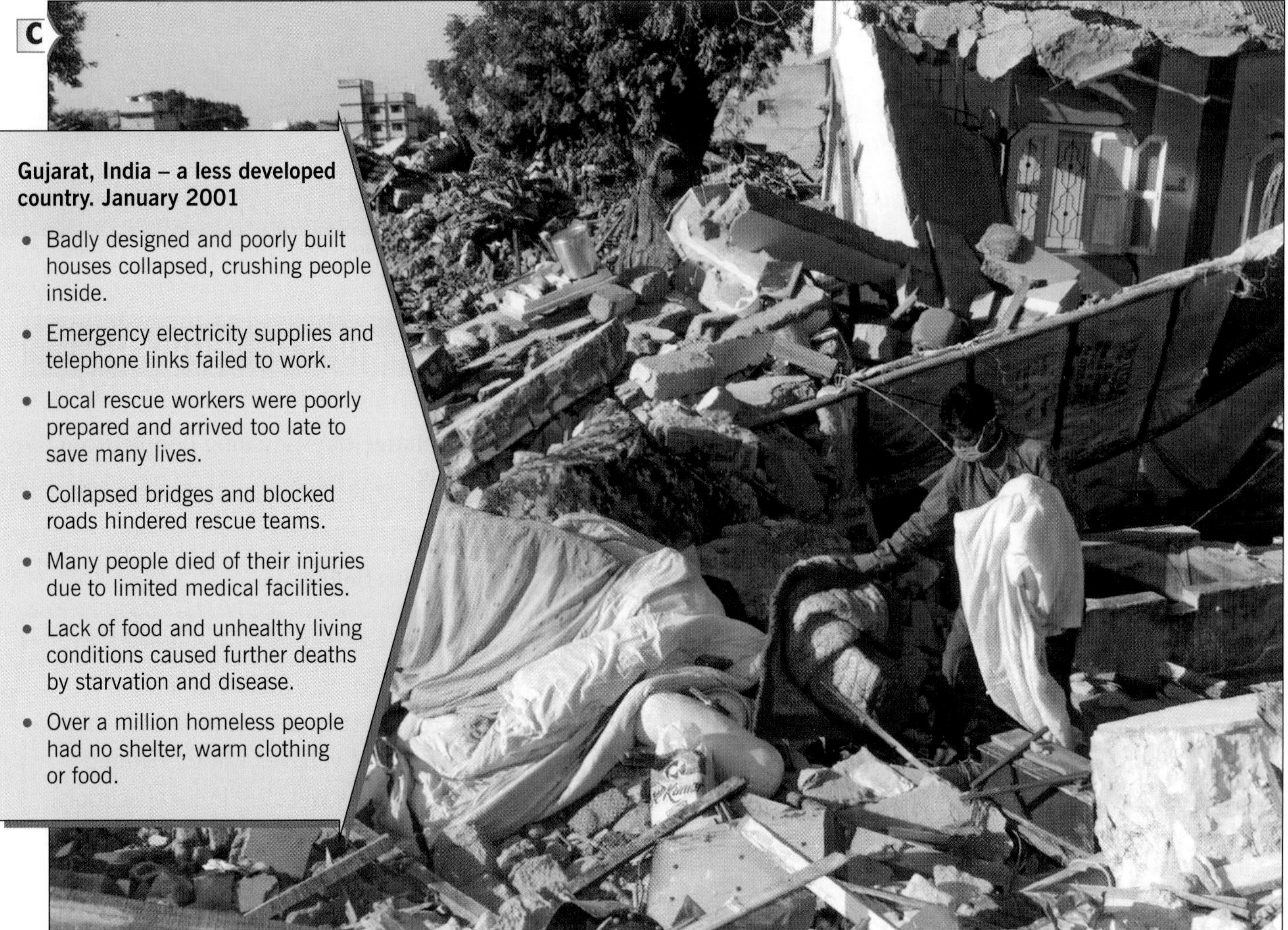

Gujarat, India – a less developed country. January 2001

- Badly designed and poorly built houses collapsed, crushing people inside.
- Emergency electricity supplies and telephone links failed to work.
- Local rescue workers were poorly prepared and arrived too late to save many lives.
- Collapsed bridges and blocked roads hindered rescue teams.
- Many people died of their injuries due to limited medical facilities.
- Lack of food and unhealthy living conditions caused further deaths by starvation and disease.
- Over a million homeless people had no shelter, warm clothing or food.

The volcano enquiry

Volcanic eruptions can cause serious problems. They may put people in danger and cause severe damage to property and the surroundings. People in the Mount Etna area are well used to their volcano erupting. They can never stop the eruptions but have learned ways of reducing the damage and danger that the eruptions cause.

In this enquiry you should imagine that you are a member of an eruption emergency team based in Catania, close to Mount Etna. The volcano has erupted and a huge lava flow is threatening the surrounding area. Over the last two weeks the eruptions have got worse and lava is expected to flow for several months and travel large distances.

You have been called to a team meeting to discuss the eruption and to decide what action to take. The lava is coming from a new vent close to the central crater (marked X on drawing **C**). Your engineers think that they can change the direction of the lava flow by using explosives and building dams.

Your job is in two parts. First you have to decide which way to divert the lava so that it causes the least danger and damage. Then you must implement an evacuation plan for those people who are in most danger.

A Mount Etna lava flow

How can the damaging effect of a volcanic eruption be reduced?

1 **a** Copy table **D** which shows some factors that have to be considered when choosing the best route for the lava.

b Look carefully at the drawing of Mount Etna on the opposite page. For the first factor in the table give a score for each route. Do the same for each of the other factors.

c Add up the scores in each column.

2 **a** Which route would you choose? The one with the highest score should be the best.

b Describe the route that you have chosen.

c Describe the advantages and any disadvantages of the route.

3 You are in charge of evacuating people from the danger zone.

a What does 'evacuate' mean?

b Complete the evacuation plan by answering the questions below.

B

Mount Etna Emergency Evacuation Plan

- Who needs to be evacuated first?
- What should they take with them?
- What types of vehicles will be needed?
- Many evacuees will stay in local schools. What items will be needed to make them comfortable?

C

The Mount Etna area

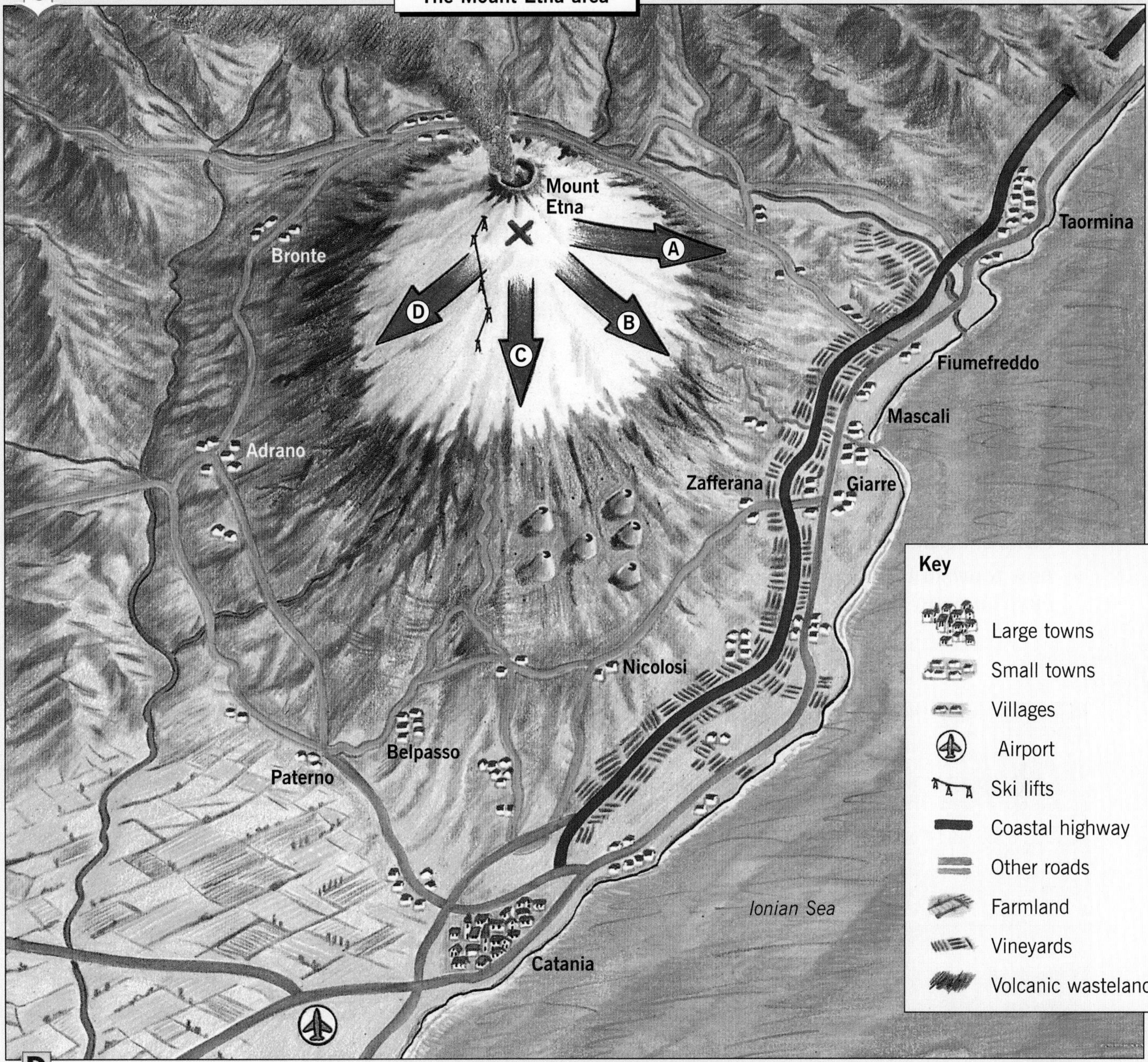

Mount Etna
A
B
C
D
Bronte
Adrano
Paterno
Belpasso
Nicolosi
Catania
Zafferana
Giarre
Mascali
Fiumefreddo
Taormina
Ionian Sea

Key

- Large towns
- Small towns
- Villages
- Airport
- Ski lifts
- Coastal highway
- Other roads
- Farmland
- Vineyards
- Volcanic wasteland

D

Factors affecting route choice	Route A	Route B	Route C	Route D
Misses all large towns				
Misses most small towns and villages				
Avoids main skiing facilities				
Misses coastal highway				
Affects few minor roads				
Misses Catania airport				
Does little damage to vineyards				
Spreads over farmland causing least damage				
Spreads harmlessly over wasteland				
Total				

Score **1** to **5** for each route.

5 if route is **excellent**

4 if route is **very good**

3 if route is **good**

2 if route is **poor**

1 if route is **unsatisfactory**

3 Tourism

What is tourism?

What is this unit about?

This unit explains how the development of tourism may bring benefits but can also cause problems. It shows how careful planning is needed to maximise the benefits and minimise the problems. It also looks at how both physical and human factors can affect the development of tourism.

In this unit you will learn about:

- **the tourist industry**
- **the problems tourism can cause**
- **management and conflict in National Parks**
- **different types of holiday**
- **how tourism can change the environment.**

Why is learning about tourism important?

We all like holidays – it is a time when we can relax and enjoy ourselves. Learning about tourism can help make those holidays even more enjoyable and interesting. It can also help us understand the tourist industry, an industry that creates jobs and wealth but needs careful management if it is not to cause problems for people and spoil the environment.

Learning about tourism can help you:

- choose where to go on holiday
- have a more enjoyable holiday
- know about jobs in tourism
- appreciate the need to look after tourist environments.

A Waikiki Beach, Hawaii

B Switzerland

C Yosemite National Park, USA

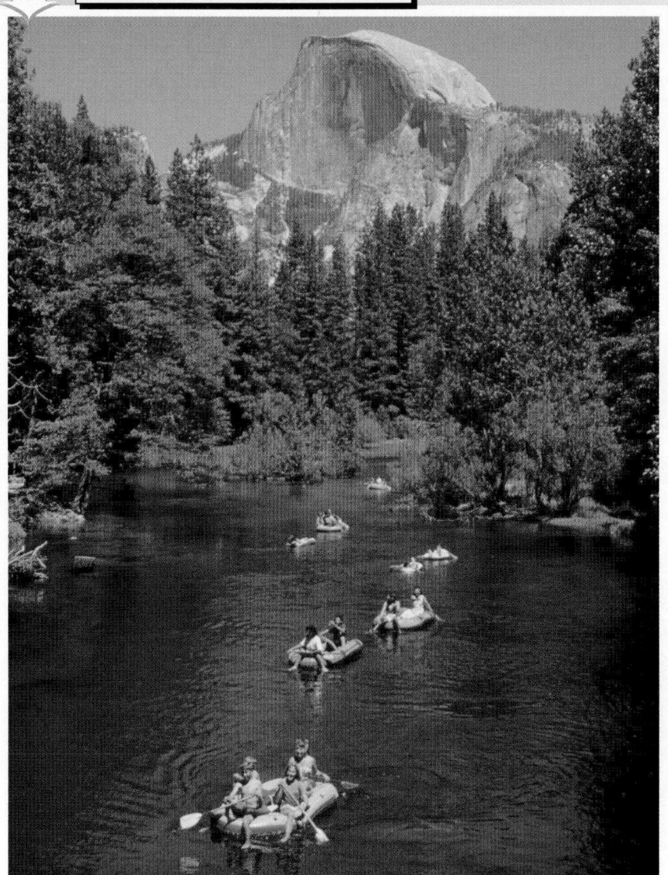

- For photos **A**, **B** and **C**:
  - describe the natural scenery
  - describe how tourism has changed the area.
- Which of the four places would you:
  - most like to visit
  - least like to visit
  - go to for a quiet time
  - go to for a busy time?

  Give reasons for your answers.

D Lake District, England

What is the tourist industry?

There are many different types of jobs. **Primary occupations** such as farming and mining involve people in collecting raw materials. **Secondary industries** employ people to make things, usually in a factory. A third type of employment is in **tertiary industries**. Tertiary industries provide a **service**. People who give help to others, such as teachers, nurses and shop assistants, are part of this industry. The number of jobs available in the tertiary sector is increasing. This is partly due to the growth of tourism.

Tourists are people who travel for pleasure. The tourist industry looks after the needs of tourists and provides the things that help them get to places to relax and enjoy themselves. The industry employs a large number of people. Travel agents, hotel waiters, tour guides and other such people rely on tourism for their livelihood. Sketch **B** shows some occupations in the tourist industry. Can you think of any more?

Tourism is big business. It is one of the world's fastest growing industries and now employs more people worldwide than any other industry. In the year 2005, £24.8 billion was spent on tourism in Britain alone. Graph **A** shows the increase in tourist numbers worldwide.

There are no signs of this increase slowing down. Can you estimate from the graph how many tourists there may be by the year 2010?

The development of the tourist industry can bring many benefits. For tourists it can improve the chances of having a good holiday. For the areas and countries involved it can be an important source of money and employment. Places like Spain, Greece and Italy that were once very poor now have much higher standards of living due to increased tourism. Countries in the poorer developing world have followed their lead. Kenya, Egypt, India and the islands of the Caribbean, for example, have used money from tourism to improve their quality of life by building new schools, hospitals, roads and factories. Some of the money has also been spent on further developing the tourist industry.

Some of the benefits that the tourist industry may bring are shown on the suitcase labels below.

C

Activities

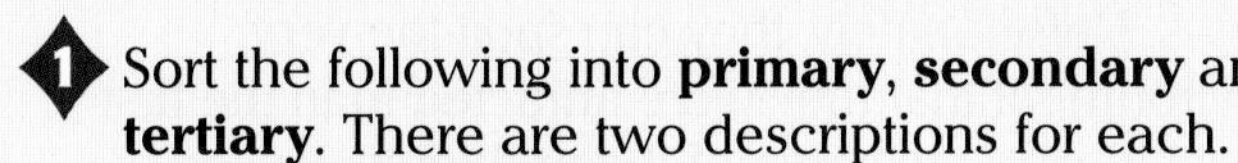

1. Sort the following into **primary**, **secondary** and **tertiary**. There are two descriptions for each.
   - Provides a service
   - Something is made
   - Provides raw materials
   - Usually done in a factory
   - Product comes from the land or sea
   - Is helpful to people
2. Write down four facts about the tourist industry.
3. Which of the jobs in sketch **B** would you most like to do? Give reasons.
4. Make a list of all the different jobs that might be found in a large tourist hotel. Try to put down more than fifteen.
5. List the jobs of people in the tourist industry who have helped you at some time. Say where and when each one happened.
6. Tourist developments can help both tourists and local people. Make a copy of table **D** and sort the statements from drawing **C** into the correct columns. Some statements will appear in both columns.

D

Tourists	Local people

Summary

Tertiary industries provide a service for people. People who work in the tourist industry are part of the tertiary sector. Tourism is one of the fastest growing industries in the world today. It can bring wealth, help create jobs and provide improved facilities for local people.

What problems does tourism cause?

Over the last thirty years or so there has been a rapid increase in tourism. More people are taking holidays, many are having several holidays a year, and travel abroad is becoming increasingly popular. There are three main reasons for this. The first is that most people are now better off than in the past and have more money available for luxuries such as holidays. Secondly, people have more leisure time and the length of their annual holidays has increased. Thirdly, places have become more accessible as transport improvements have made travel faster, easier and cheaper.

This increase in tourism has brought many benefits but it has also caused problems. Photos **A** and **B** show two of these problems – attractive places spoilt by rubbish left by tourists in one case, and overcrowding in the other. Many places that hoped to gain from tourism have been disappointed. Some of the reasons for this disappointment are given below.

- The better tourist jobs rarely go to local people.
- Jobs are seasonal so there is no work for much of the year.
- Tourism raises prices so locals cannot afford goods in shops.
- Local people cannot afford to use tourist facilities.
- Discos, bars and other tourist attractions spoil the local way of life.
- Most of the money from tourism goes out of the area.

B An overcrowded tourist resort

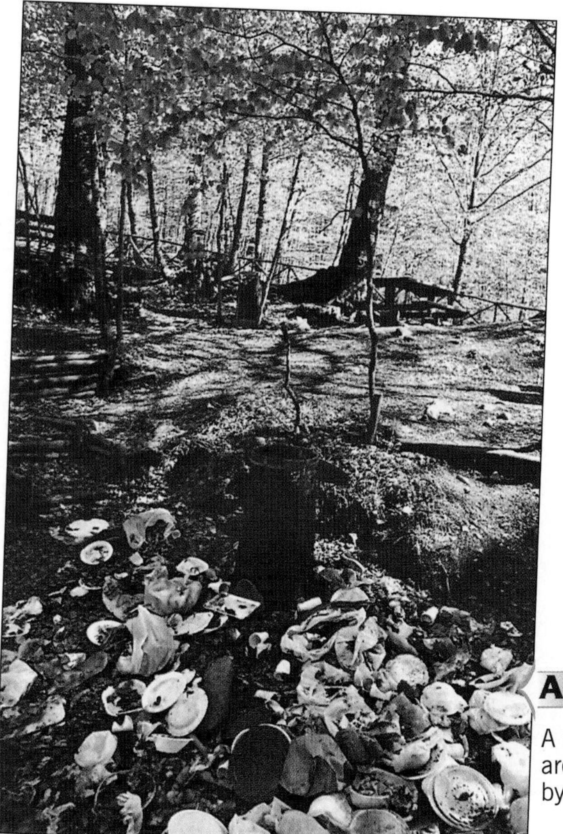

A A tourist area spoilt by litter

Some of the worst problems occur in the countryside. People go there for peace and quiet, and to enjoy the views. Unfortunately, they can spoil the very environment that they were attracted to in the first place. Below are some of the problems caused by tourism in the countryside. See if you can spot them and any others on cartoon **C**.

- Narrow country roads are blocked by traffic.
- Attractive landscapes are spoilt by tourist buildings.
- Litter looks unsightly and is a danger to animals.
- Walls are knocked down by careless tourists.
- Gates are left open allowing animals to get out.
- Popular locations are overcrowded and spoilt.
- Farming land is damaged.
- Farmers are unable to go about their business.
- Wildlife is frightened away.
- Trees and plants are damaged.

Tourism will not go away. People will always want to have holidays. Holiday areas will always want to have tourists around. What we need to do is to plan and manage the tourist industry carefully. We must try to increase the good effects but reduce the bad ones.

C

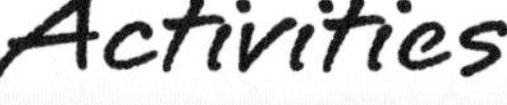

Activities

1. Give three reasons for the increase in tourism.

2. The results of tourism can be disappointing for some places. Six reasons for this are given on page 50. Copy table **D** and put each reason in the correct column.

3. **a** Match each of the labels from drawing **E** with the correct number in cartoon **C**.
   b Which four do you think would cause most problems for farmers in the area? Give reasons for each of your answers.
   c Which four do you think would most spoil the area for tourists? Give reasons for each of your answers.

D

Money	Jobs	Others

E

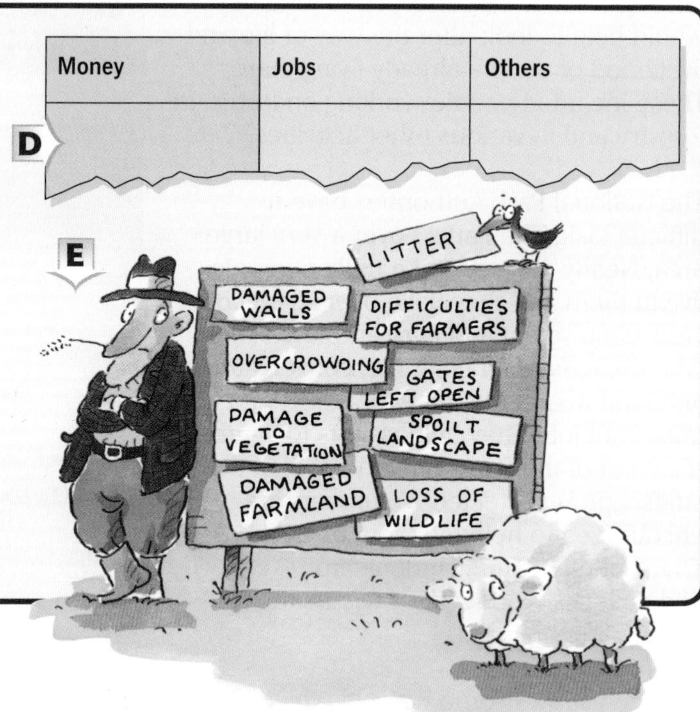

EXTRA

Choose any two of the problems shown in cartoon **C**. For each one suggest what might be done to prevent it or at least reduce its bad effects.

Summary

Tourist activities can bring enjoyment and create employment and wealth. In some places tourism can cause problems for people and spoil the environment. Care is needed in planning and managing tourist environments.

What are National Parks?

National Parks are large areas of beautiful countryside. Their scenery and wildlife are protected so that everyone can enjoy them. The world's first Park was opened as long ago as 1872 at Yellowstone in the United States. Since then many countries have set up similar Parks. Parks around the world vary in many ways but they all have the same two aims.

1 To preserve and care for the environment.
2 To provide a place for recreation and enjoyment.

Britain's first National Parks were set up in the 1950s. At that time the government felt there was a real danger that some of Britain's finest scenery would be damaged or permanently destroyed. The main idea of the Parks was to provide protection for the environment. It was also hoped that they would help to look after the way of life and livelihood of people already living there. These included people working on farms, in forestry and in various other activities.

The National Park Authorities have a difficult task. The Parks cover a very large area. Nearly a quarter of a million people live in them and they have over 90 million visits each year. Each Park employs staff who are experts in planning, conservation and land management. Park rangers do an important job. They help visitors to get the most out of the Parks and ensure that the landscape is protected. Volunteers are encouraged to help look after the Parks. These often include students from schools and colleges who work at weekends or during their holidays.

Sketch **A** shows some ways in which the Parks protect the countryside and help local people and visitors.

A

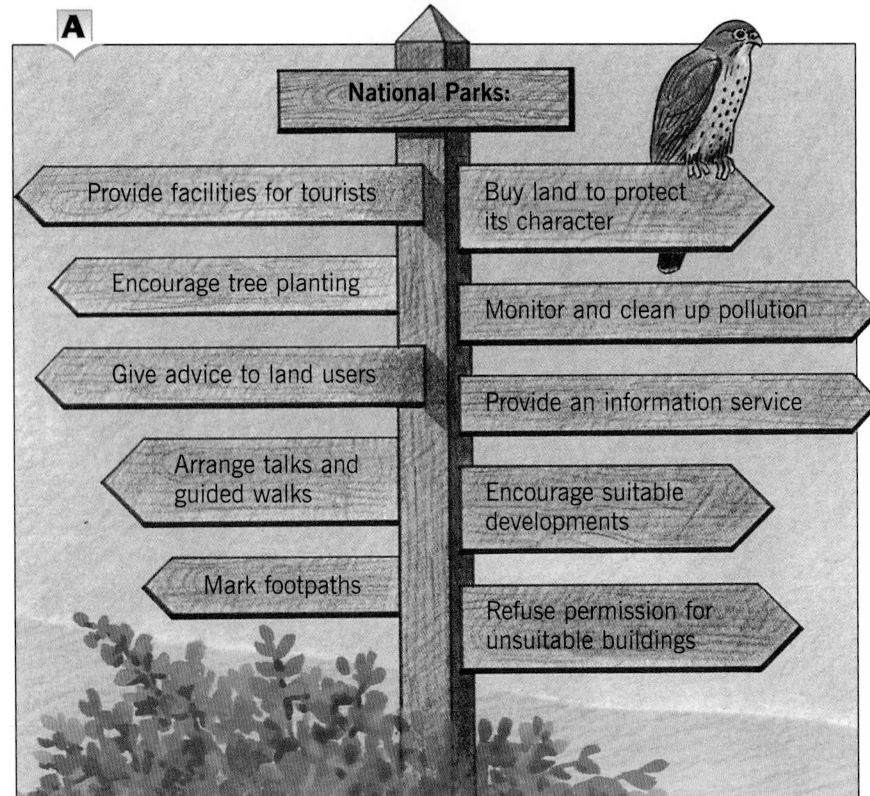

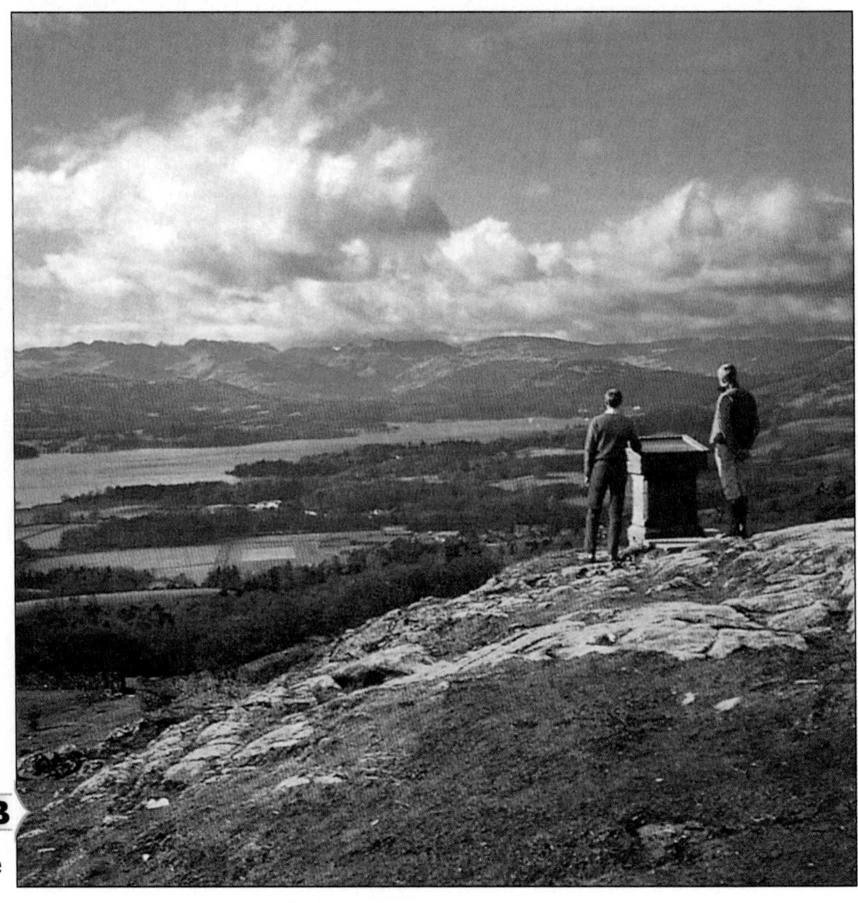

B

Orrest Head, Windermere

C National Parks in England and Wales

Key
- Original ten National Parks
- Recent or planned National Parks
- Highland
- Areas with most people
- ● Main towns
- 0 – 100 km

Northumberland
Newcastle upon Tyne
Lake District
North York Moors
Yorkshire Dales
Leeds
Manchester
Liverpool
Sheffield
Snowdonia
Peak District
Birmingham
The Broads
Brecon Beacons
Pembrokeshire Coast
Cardiff
London
Exmoor
Southampton
South Downs
Dartmoor
New Forest
N

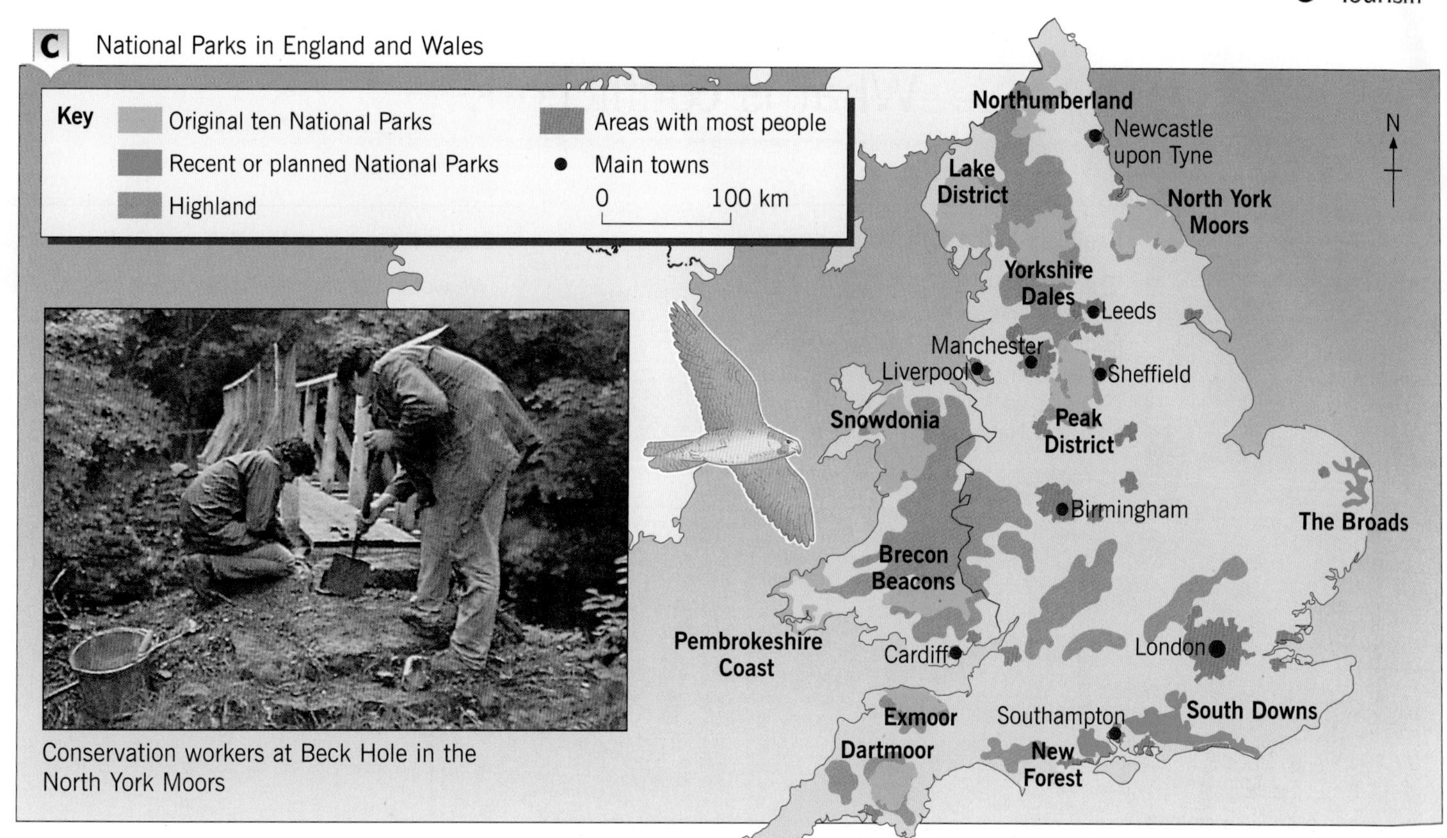

Conservation workers at Beck Hole in the North York Moors

Activities

1 **a** Name the original National Parks.

b Which Park is nearest to where you live? How far away is it?

b How many of the Parks are in the highland areas?

2 **a** How many National Parks are in Wales?

b Until recently, which large area of population was furthest from any Parks?

c Which highland Park is likely to have most visitors? Give a reason.

3 What are National Parks?

4 Look at sketch **A**. Give four ways in which National Park Authorities help visitors.

EXTRA

1 Sketch **D** is a simplified drawing of photo **B**.

a Make a larger copy of the sketch.

b Add colour to make it clearer and more attractive.

c Label and mark with arrows the following features.

mountains | hills | lake | forest | farmland | houses | park information

2 Photo **B** is of a typical National Park. Write to a friend, describing the Park. Try to include these words in your letter:
hilly, beautiful, quiet, peaceful, natural, lovely, attractive, relaxing.

D

Summary

National Parks help to protect our countryside. National Park Authorities care for the landscape and help to look after the needs of visitors and the people who live and work there.

What is conflict?

The page opposite shows some of the problems people have in National Parks. These problems can cause **conflict**. Conflict is when there is disagreement over how something should be used.

Land ownership is a main cause of conflict in the Parks. As graph **A** shows, very little of the land is owned by the nation for everyone to use and enjoy. Most of it belongs to private individuals like farmers and house owners. Farmers may not want tourists to walk across their land. House owners may not like visitors parking outside their houses and blocking drives or leaving litter.

An important job of the National Park rangers is to try to reduce this conflict for the benefit of all concerned.

A Land ownership in Britain's National Parks

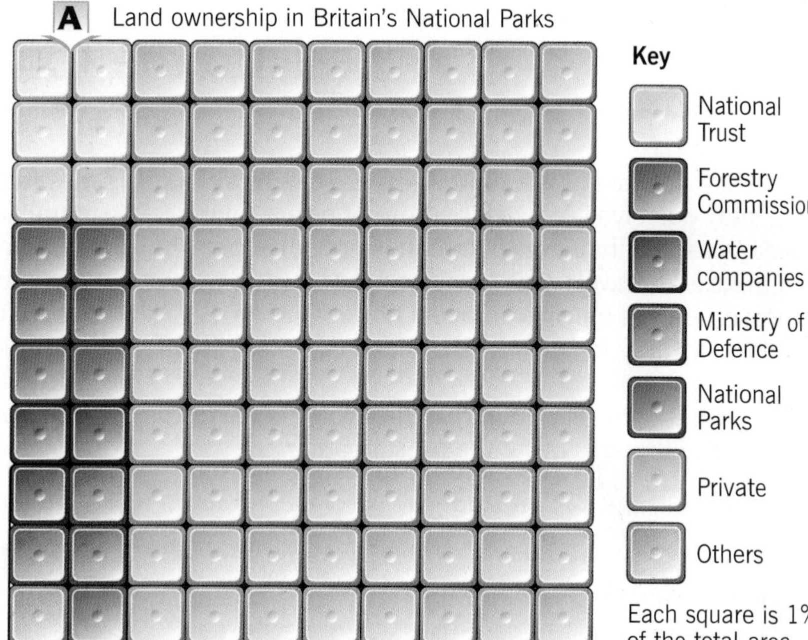

Activities

1. What is conflict?
2. Conflict arises in National Parks between people in many ways. The pictures in **B** show some of these people. Match up the pictures with the statements numbered **1** to **6**.
3. Look at graph **A**.
   - **a** What percentage of land is privately owned?
   - **b** How much land is owned by National Parks?
   - **c** Who are the second largest land owners?
   - **d** Which land owner may cause most conflict with tourists? Give reasons.

B

EXTRA

You will need to work with someone else for this activity. Discuss whether the following should be allowed in National Parks. Write down reasons for your answers.

1. Motor cycle scrambling
2. Motor boat racing
3. Mountain bikes
4. A motorway
5. A leisure park

Summary

The needs of different people can cause conflict in areas of attractive countryside. Careful management by National Park Authorities can help reduce the problem.

Where do the tourists go?

We all like holidays. It is a time when we can relax and enjoy ourselves. Nowadays more of us are taking holidays abroad than ever before. It is now cheaper and quicker to travel than it used to be, and improvements in transport make it possible to go almost anywhere in the world.

Whilst just about every country in the world is visited by tourists, Europe is by far the most popular destination. As table **A** and map **B** show, Europe attracts more than half of all the world's tourists and has six of the world's top ten tourist destinations. The most popular areas are the countries of the Mediterranean where hot sun and blue skies are common throughout the summer (pages 16 and 17). The availability of cheap flights also makes holidays in Europe very affordable.

A Top ten tourist destinations, 2004

Country	Tourist arrivals (millions)	Tourist earnings (£ billions)
1 France	75.1	21.5
2 Spain	52.3	24.5
3 USA	40.4	38.3
4 Italy	39.6	18.4
5 China	33.0	10.2
6 UK	24.8	11.4
7 Austria	19.1	8.0
8 Mexico	18.7	7.6
9 Germany	18.4	13.5
10 Canada	17.5	12.1

B International tourist arrivals, 2004

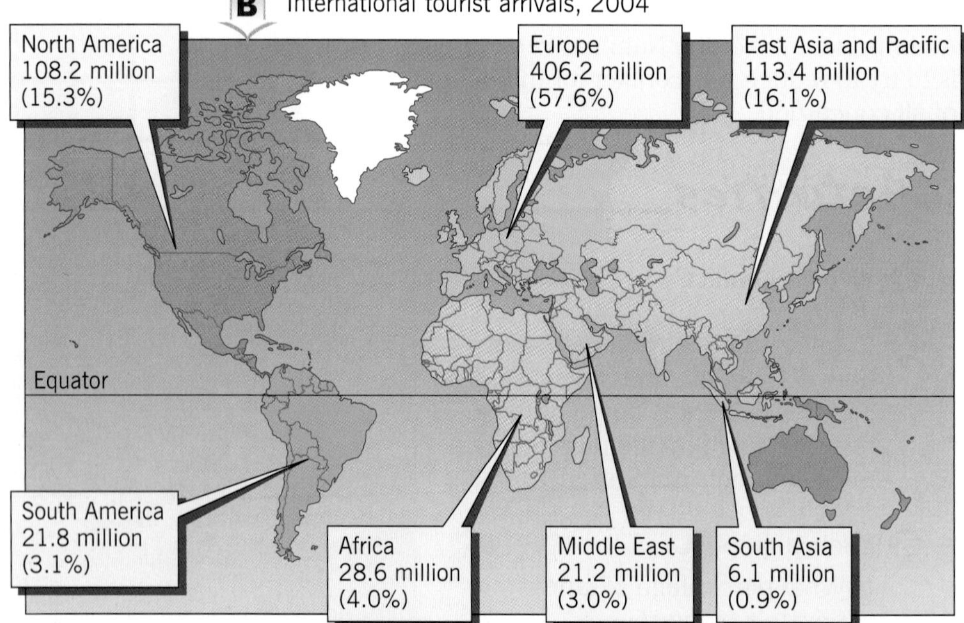

As well as there being more places to visit around the world, there is now a greater variety of holidays available to tourists. Some of these are shown in drawing **D** on the opposite page. Others include adventure, activity and interest holidays. Ocean cruises, city breaks and even shopping holidays are also popular.

So given this great choice of holiday, how do we decide where to go? Most of us are influenced by the adverts that we see but there is more to it than that. We must learn to ask questions and try to find out more about a place before we make a choice. Drawing **C** shows some questions to think about when choosing a holiday.

C

D

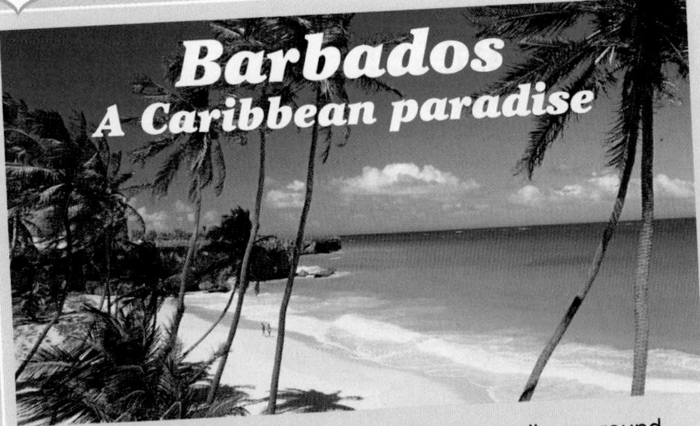

Barbados offers you hot sun and warm sea all year round. There are fine sandy beaches, swaying palm trees, lovely hotels and fine restaurants. Have a relaxing beach holiday or enjoy excellent watersports activities.

Prices from £1,260 for two weeks

Kenya has it all – great weather, spectacular scenery and exciting wildlife. Enjoy a safari and see lion, elephant, giraffe, zebra and many more. Then soak up the sun and laze on the unspoilt beaches of the Indian Ocean.

Florida

Glorious sunshine and hot all year. The wonderful world of Disney, the Epcot Centre, Universal Studios and Wet 'n' Wild. Fun and entertainment for all the family. Then chill out on the beach and maybe swim with the dolphins.

Come to Switzerland's top ski resort. See the famous Matterhorn and use some of Europe's best winter sports facilities. Enjoy the delights of a friendly and interesting alpine village.

Two weeks from £620

Activities

1. Look at map **B**.
   - **a** List the regions in order of popularity.
   - **b** Which region was by far the most visited?
   - **c** Suggest as many reasons as you can to explain this popularity.
2. List the attractions of Kenya as a holiday destination. Try to give at least six.
3. Look at the people shown in drawing **E**. Choose the most suitable holiday for them. Give reasons for your choice.
4. Given a choice, where in the world would you like to go and what would you do? Give reasons for your answer.

We've got three children. We need somewhere with plenty for them to do and where we can relax.

A young family

We would like a winter holiday where it is quiet and warm.

A retired couple

Summary

Cheaper and easier travel has made it possible to holiday almost anywhere in the world. Deciding where to go needs careful thought.

Mallorca – a holiday paradise?

A

0 5 10 15 km
N
France
Spain
Portugal
Mallorca
Mediterranean Sea
Africa

Key
Main road
Minor road
Railway
Ferry
Airport
Beach
Places of interest
Height in metres
500 to 1000
100 to 500
Below 100

Cape Formentor
Port de Pollença
Torrent de Parels (waterfall)
Pollença
Alcudia
Puig Major 1445 m
Monastery of Lluc
Bay of Alcudia
Puerto de Soller
Soller
Albufera wildlife refuge
Inca
Magdalena sanctuary
Cala Rajada
Cala Millor
Palma
Manacor
Port d'Andratx
Portocristo
S' Arenal
Caves of Drach
Palma Nova
Magaluf
Bay of Palma
Santueri Castle
Campos
Portocolom
Cala d'Or
To Barcelona
Cala Figuera

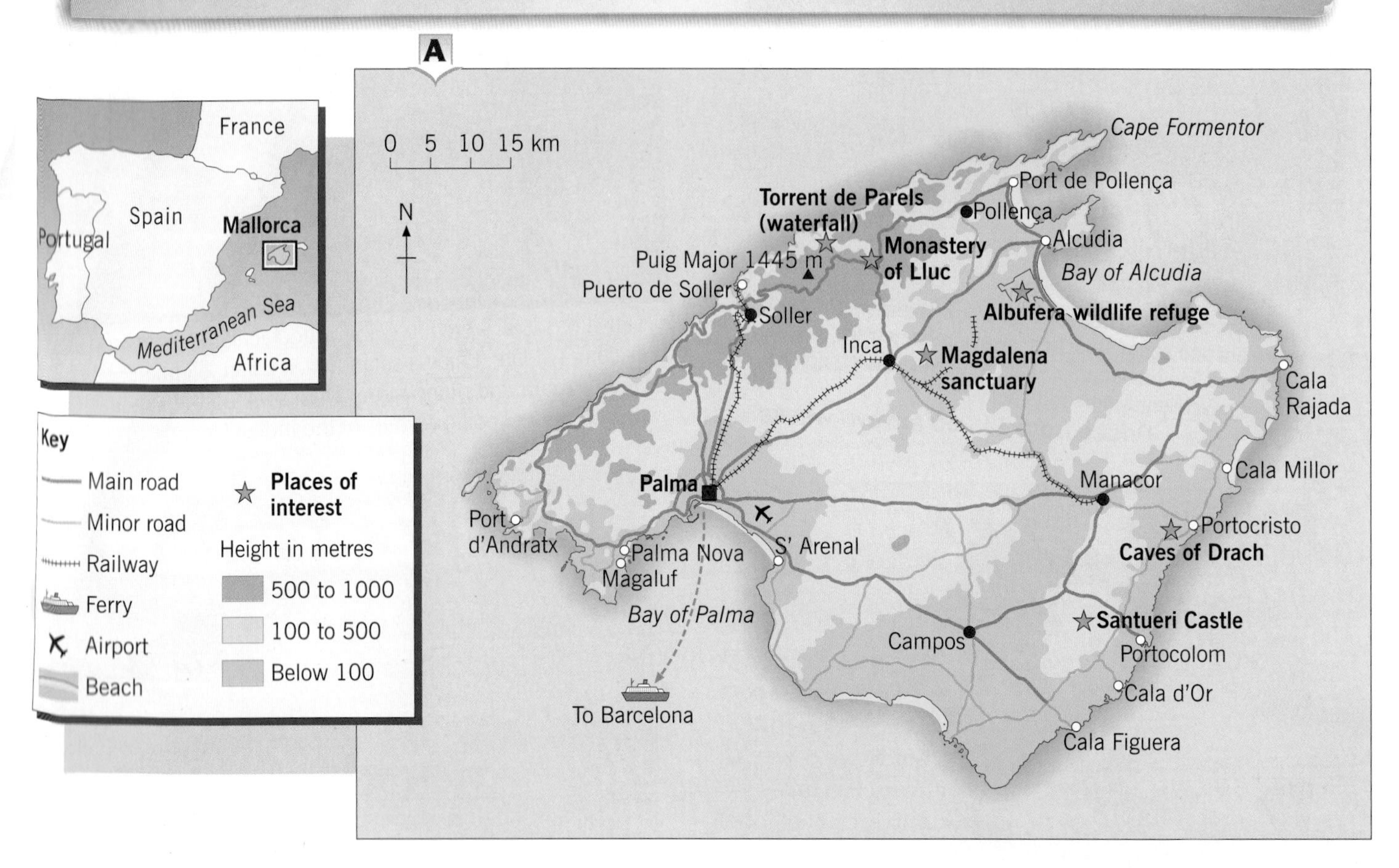

The Spanish island of Mallorca (also spelled Majorca) is the largest of the Balearic Islands and lies about 200 kilometres off the east coast of Spain. Many people consider it to be the most beautiful of all the Mediterranean islands. On the right is part of a travel brochure which describes the island.

Over 7 million tourists visit Mallorca every year. The island has a permanent population of 767,000 and measures little more than 3,640 square kilometres. It is not surprising that the effects of tourism can be seen everywhere.

Some of the effects are bad. Many islanders worry about the noise, litter, violence and drunkenness that have become common in some of the more developed areas. They also worry about the effects of tourism on Spanish culture and traditions. They argue that many of the jobs provided are unskilled, poorly paid and seasonal.

Most islanders, however, are willing to put up with the problems caused by tourism. They are grateful for any employment opportunities that come their way. They feel that the money that tourism brings to the island more than balances the damage that comes with it.

B

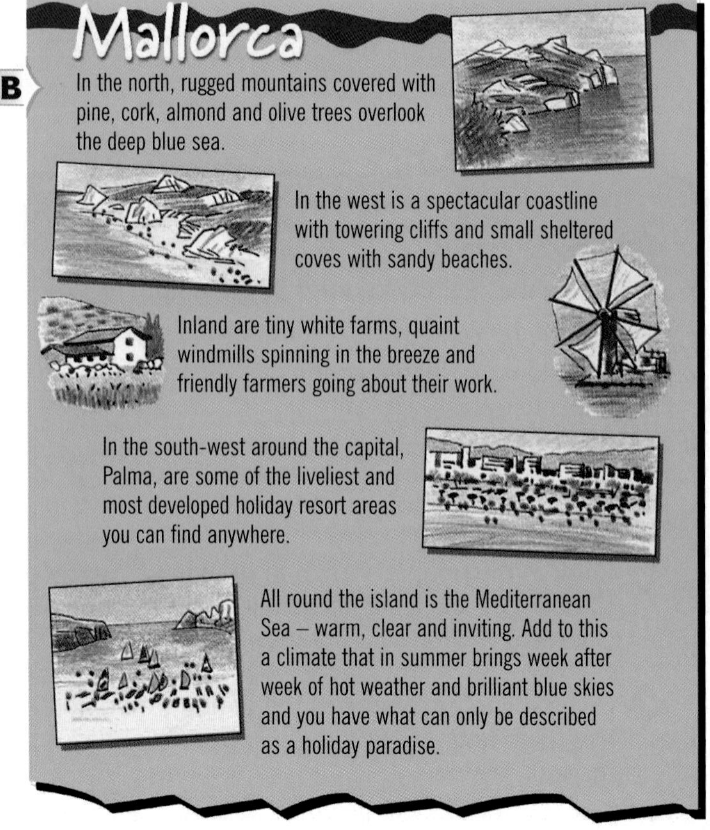

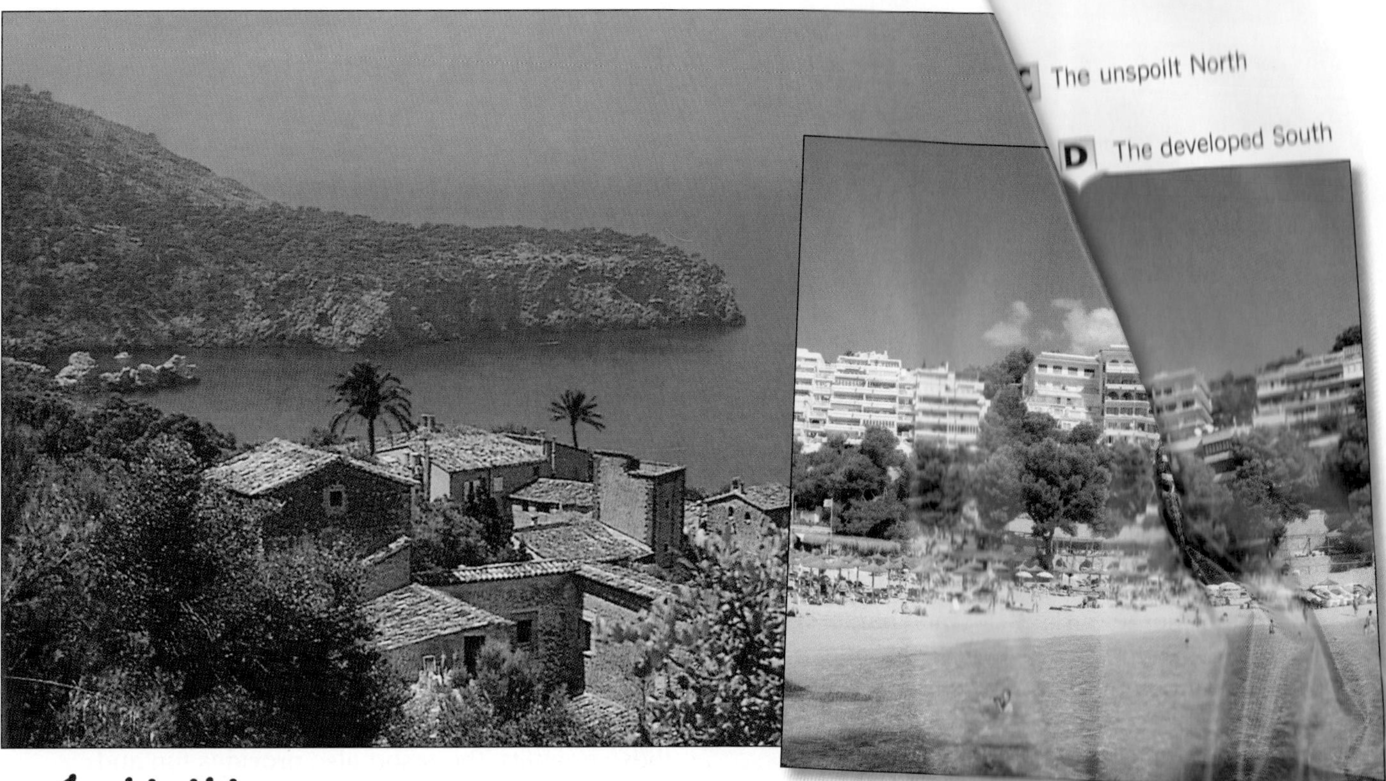

C The unspoilt North

D The developed South

Activities

1. Make a Fact File using the headings in **E** below. Add any other things about Mallorca that interest you.

E

Fact File

Name of islands ______

Distance from Spain ______

Population ______

Capital ______

Weather ______

Size ______

Physical features ______

Vegetation ______

Tourists per year ______

2. Look at photos **C** and **D**. Sort the following under the headings **Unspoilt North** and **Developed South**. Some may go under both headings.

Hot, sunny weather, many tourists, crowded, open spaces, noisy, many hotels, nightlife, lovely views, beaches, peace and quiet, farmland.

3. **a** Write down the good effects of tourism for Mallorca.
   b List at least six problems that tourism has caused in Mallorca.

4. Look carefully at map **A**.
   a Make a list of all the things that interest you and the places you would like to visit.
   b Plan how you would spend a week in Mallorca. Describe what you would do and see on each day.

EXTRA

Produce a small booklet about Mallorca. Try to include:

- a map and description of where it is
- a simple map of the island
- a page on its natural attractions
- a page on tourist developments
- a page to show the advantages and disadvantages that tourism has brought.

You will be able to get some ideas and photos from a travel brochure or on the internet.

Summary

Mallorca is the Mediterranean's most popular holiday island. Its attractions include spectacular scenery, glorious weather and lively seaside resorts. Some people are concerned about the bad effects of tourism.

D y Florida – a holiday dream?

Disney leisure resorts ar n throughout the
world. They were first d ed in the 1950s by
Walt Disney. His idea v provide fun, thrills and
adventure for people b ilding a world of dreams,
fantasies and make-b e.

The first Disneylan ned in California in 1955.
This was followed lorida's Magic Kingdom in 1971,
Tokyo Disneyland 1983, Disneyland Paris in 1992
and, most recen Dubai Disney in 2006.

The best know f the resorts is probably the Walt Disney World Resort in Florida, as it is now known. It is home to four theme parks, two water parks, six golf courses, more than twenty resort hotels and various shopping and entertainment areas. It is the largest theme park resort in the world and covers an area almost twice the size of Manchester.

Great care was taken when choosing the site for the resort. There were four main requirements.

1 There needed to be a large local **population**.
2 The area already had to be popular with **tourists**.
3 **Transport** systems and accessibility had to be good.
4 A large area of **land** had to be available.

Some of the reasons for choosing the site near Orlando are shown on map **A** below. The choice of site was a good one. The resort attracts over 15 million people a year and is the world's most visited tourist attraction.

The Disney resort has been good for Florida. It has increased tourism and brought money into the state. It has also brought about improvements in transport and created several thousand new jobs. As well as these benefits, the resort also provides fun and enjoyment for the many local people who visit the resort in their own leisure time.

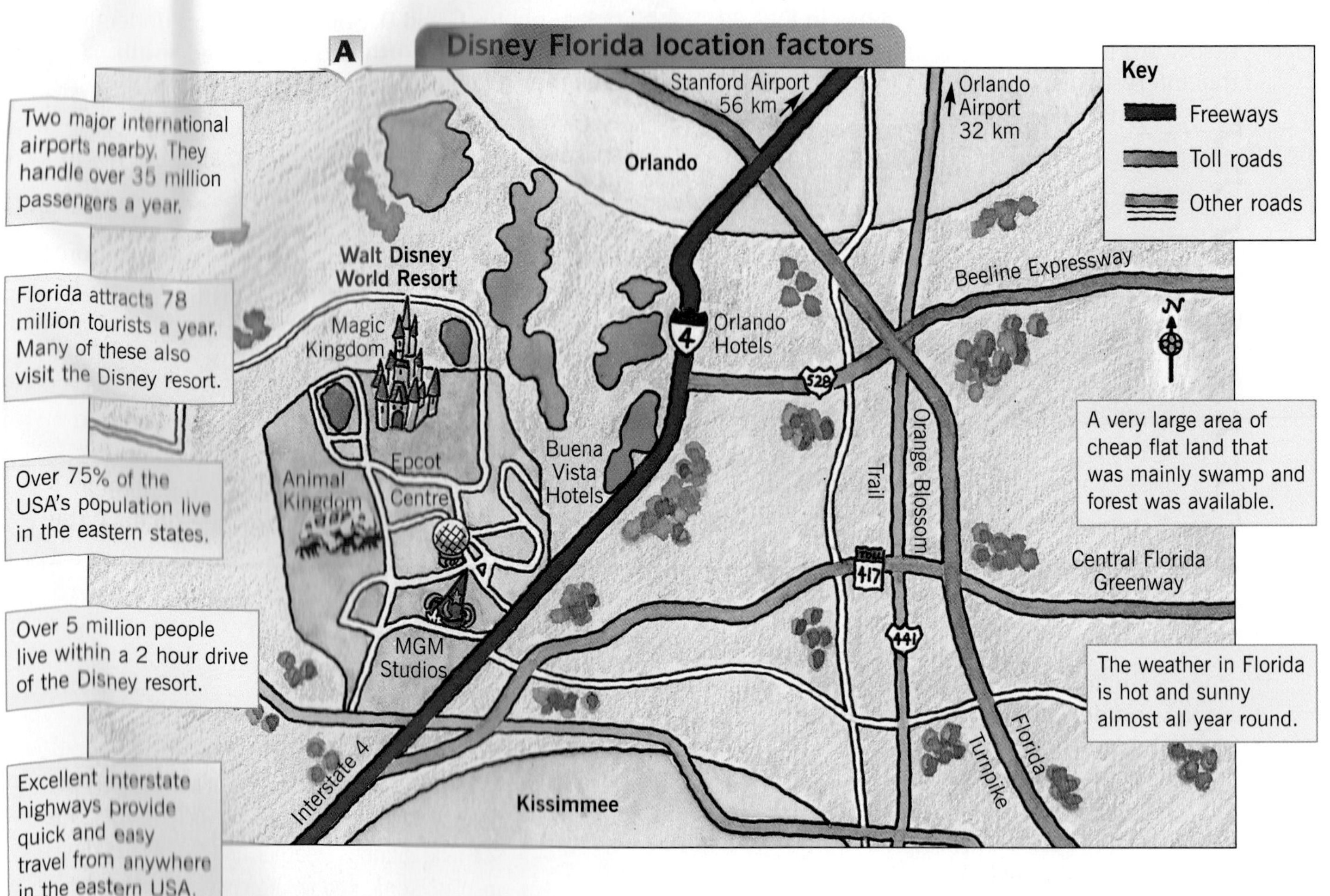

B Walt Disney World Resort in Florida

Magic Kingdom

This is the park where fairy tales come true. There is Cinderella's Castle, adventures of the Wild West and African jungle plus thrilling rides like Thunder Mountain Railroad. There are also shows, shops, parades, restaurants, themed hotels and much, much more.

The Magic Kingdom

Epcot Centre

This is Disney's idea of the future where you can experience the wonders of technology and explore the cultures of the world. You can buckle-up at Test Track, blast-off at Mission:Space, soar around the world on a glider and stroll around 11 countries in World Showcase.

Animal Kingdom

This is full of wild animal adventures, spectacular shows and more than 250 animal species. You can travel on safari and be surrounded by animals, enjoy stage shows like 'The Lion King', visit the dinosaurs and try out the frighteningly fast roller coaster – Expedition Everest.

The Epcot Centre

Disney-MGM Studios

Here you can experience thrilling shows and exciting backstage action. Plunge down 13 storeys in the Twilight Zone Tower of Terror, break all the limits on the high-speed Rock'n'Roll Coaster then enjoy the night-time fireworks show, Fantasmic.

Activities

1 **a** Where and when did the first Disneyland open?

b Name three other places with Disney resorts.

c What was Walt Disney's idea?

2 Look at map **A**.

a Name two Florida airports.

b Which freeway links the airports to the resort?

c Which route would you take if you were driving from Miami to the Disney resort?

3 Make a large copy of diagram **C**. Add examples from the Florida area for each location factor.

4 Plan a trip to the Disney resort in Florida with your family. Describe how you will get there, where you will stay and what you will do. Write a postcard home.

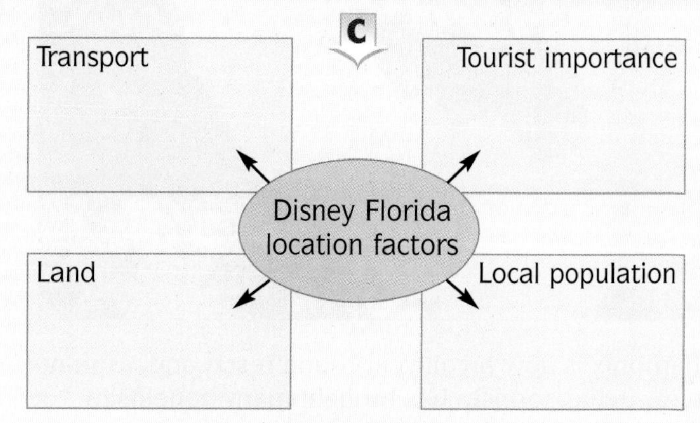

Summary

The Walt Disney World Resort in Florida has been located near Orlando because of good transport, a large population, existing tourism and suitable land.

The Alps – a winter wonderland?

Every year some 50 million people visit the Alps, two-thirds of them on winter skiing holidays. For many people, alpine resorts provide the best of everything. There is skiing, walking, climbing and sailing as well as dramatic scenery and clean, fresh air.

Chamonix is one of the most popular alpine resorts. It is located in the French Alps at the foot of Europe's highest mountain, Mont Blanc. The skiing is spectacular with 62 lifts giving access to six main ski areas. There are slopes for everyone from beginner to expert. Ski instruction is well organised and mountain restaurants cater for skiers who want to stay on the slopes all day.

B Cable cars above Chamonix

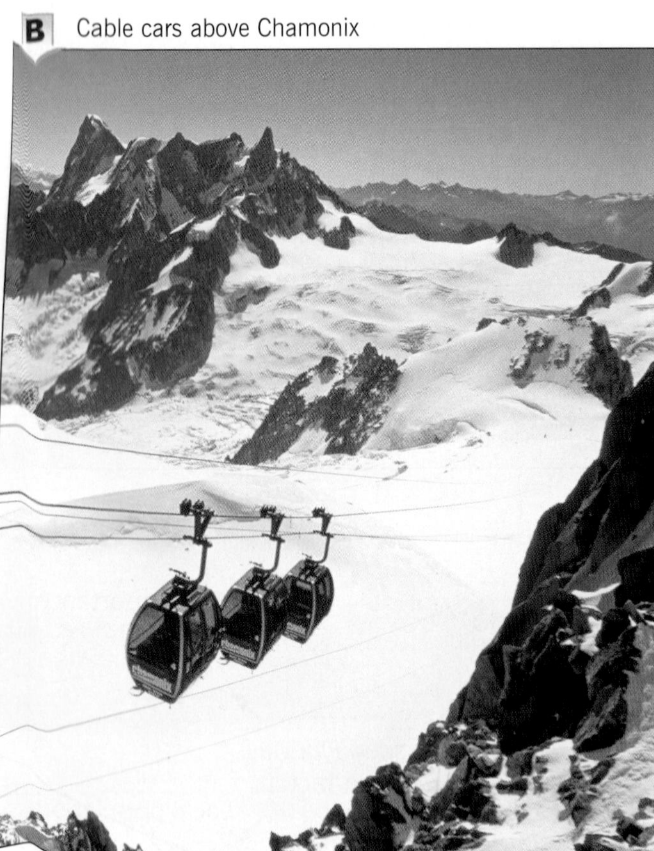

Off the slopes there is plenty to do. The town is bustling and lively and there are plenty of bars, cafés, restaurants and smart shops. For the more active there is a sports centre, large heated swimming pool, ice skating and snowmobiling. Day trips to Italy and Switzerland are also available for those who need a change.

A

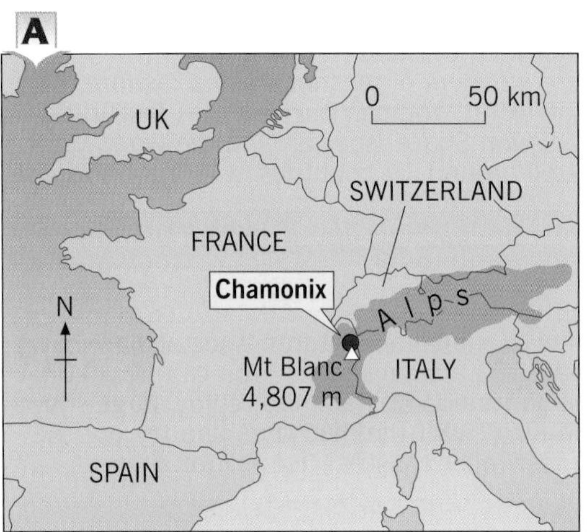

C Chamonix town

Chamonix is now an all-year-round resort and, as may be expected, tourism has brought many benefits to the town and local area. It has increased employment opportunities, raised living standards and generally improved the quality of life for most local people.

Unfortunately there have also been problems. Many people are concerned that winter sports facilities, in particular, are spoiling the countryside and permanently damaging the environment.

Chamonix, along with other mountain resorts, is concerned about these problems. The town has introduced strict new planning controls, and the emphasis is now on improving existing facilities rather than developing new ones.

Drawing **D** shows some of the effects of tourism on winter resorts like Chamonix.

D ## Winter sports – the good and the bad

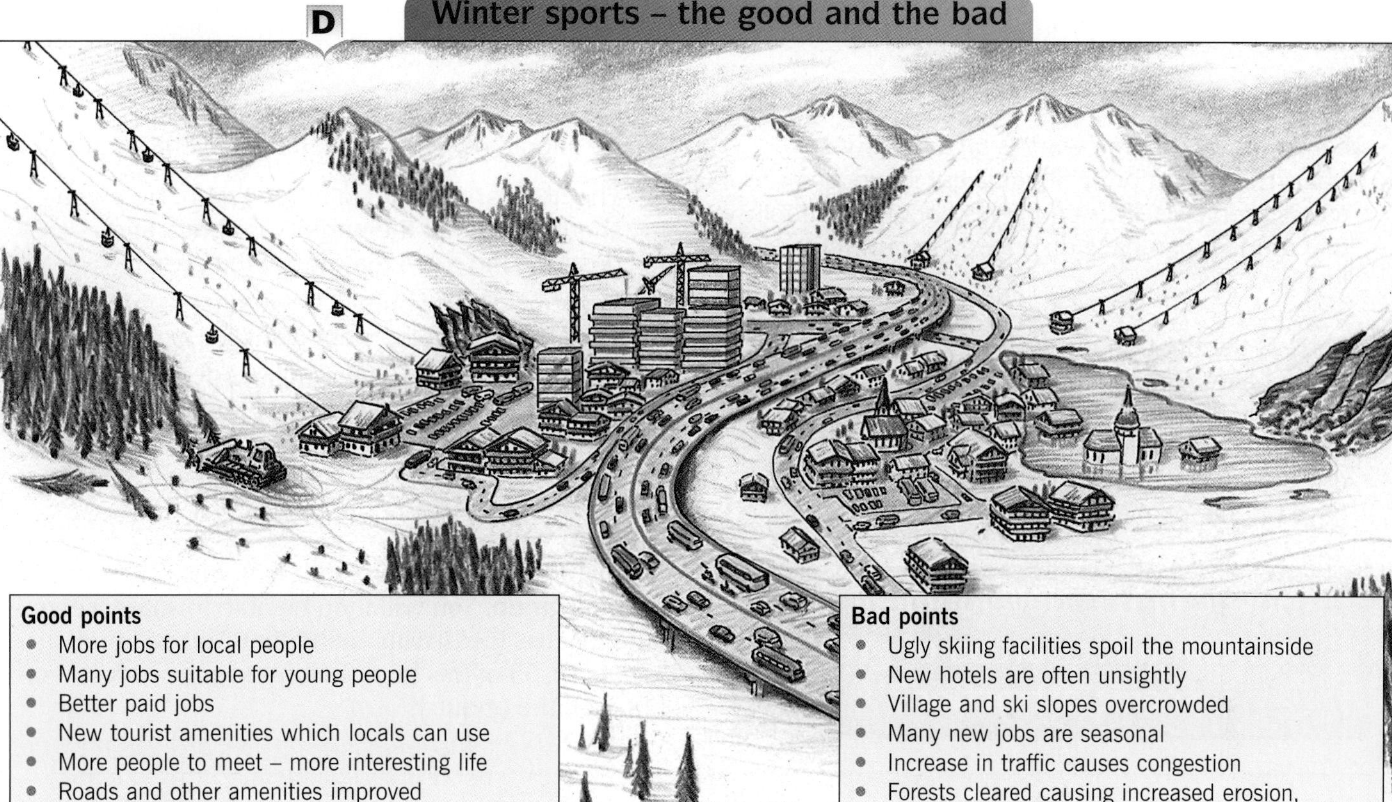

Good points
- More jobs for local people
- Many jobs suitable for young people
- Better paid jobs
- New tourist amenities which locals can use
- More people to meet – more interesting life
- Roads and other amenities improved
- Better quality of life
- Fewer people leave the area

Bad points
- Ugly skiing facilities spoil the mountainside
- New hotels are often unsightly
- Village and ski slopes overcrowded
- Many new jobs are seasonal
- Increase in traffic causes congestion
- Forests cleared causing increased erosion, landslides and flooding
- Plant life damaged and wildlife frightened away

Activities

 1 **a** In which countries are the Alps?
b How far is it from London to Chamonix?

 2 The main attractions of Chamonix are:
- fine scenery
- mountains
- entertainments
- wildlife
- wild flowers
- skiing area
- restaurants
- outdoor activities
- attractive village
- shops

List which of these you can see in photos **B** and **C**.

 3 Make a list of tourist industry jobs that may be available in Chamonix. You should write down at least eight, but try to make a really long list.

 4 **a** Make a copy of table **E**.
b Tick the correct boxes for each effect of tourism.
c Colour the benefits in green and the problems in red.
d Choose any one problem and suggest what could be done to reduce it.

E

Effect	People	Environment
New jobs		
Traffic jams		
Overcrowding		
Things to do		
Loss of wildlife		
People to meet		
Soil erosion		
Flooding		
Increased wealth		

5 Design a page for a travel brochure advertising Chamonix as a ski resort. Include information on location, skiing and other attractions. Make your page as interesting, colourful and attractive as possible.

Summary

Mountain areas are becoming increasingly popular with tourists. Europe's alpine resorts attract large numbers of winter visitors. Care needs to be taken to protect the surroundings.

The tourism enquiry

Tourism is one of the world's fastest growing industries. It can bring benefits but it can also cause problems. This enquiry is about the effects that tourism can have on an area in northern England.

Look at the photo opposite which shows a valley in the Lake District National Park. A holiday company has shown interest in developing tourism in the valley. Your task is to plan these developments, and suggest what effects they will have on the local area.

How can the development of tourism affect areas of great scenic attraction?

There should be three main parts to your enquiry.

- The first part will be an introduction. Here you should say what the enquiry is about, and describe the main features of the area.
- In the next part you will need to describe and explain the development plans for the valley.
- Finally you will need to outline the good points and the bad points of the scheme, and give your considered opinion on it.

It will be best if you can work with a partner or in a small group. You will then be able to share views and discuss ideas with each other. You will find pages 48–55 of this book helpful as you work through the enquiry.

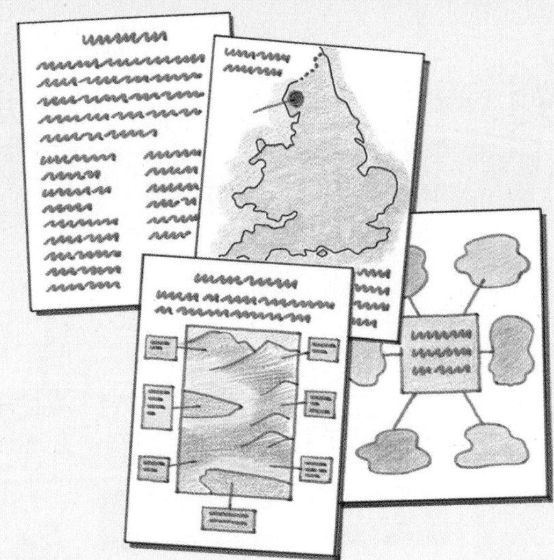

1 Introduction – what is the enquiry about?

You could use maps, sketches, lists, writing or star diagrams here.

a First look carefully at the enquiry question and the guidelines given at the top of page 64. Say briefly what you have to do and how you are going to do it.

b Next show where the Lake District National Park is located. The map on page 53 will help you.

c Now describe the main features of the valley as it is today. Use the photo below and information in the drawing opposite. A labelled sketch could be useful here.

d Explain why the valley seems to be a good place for tourist development.

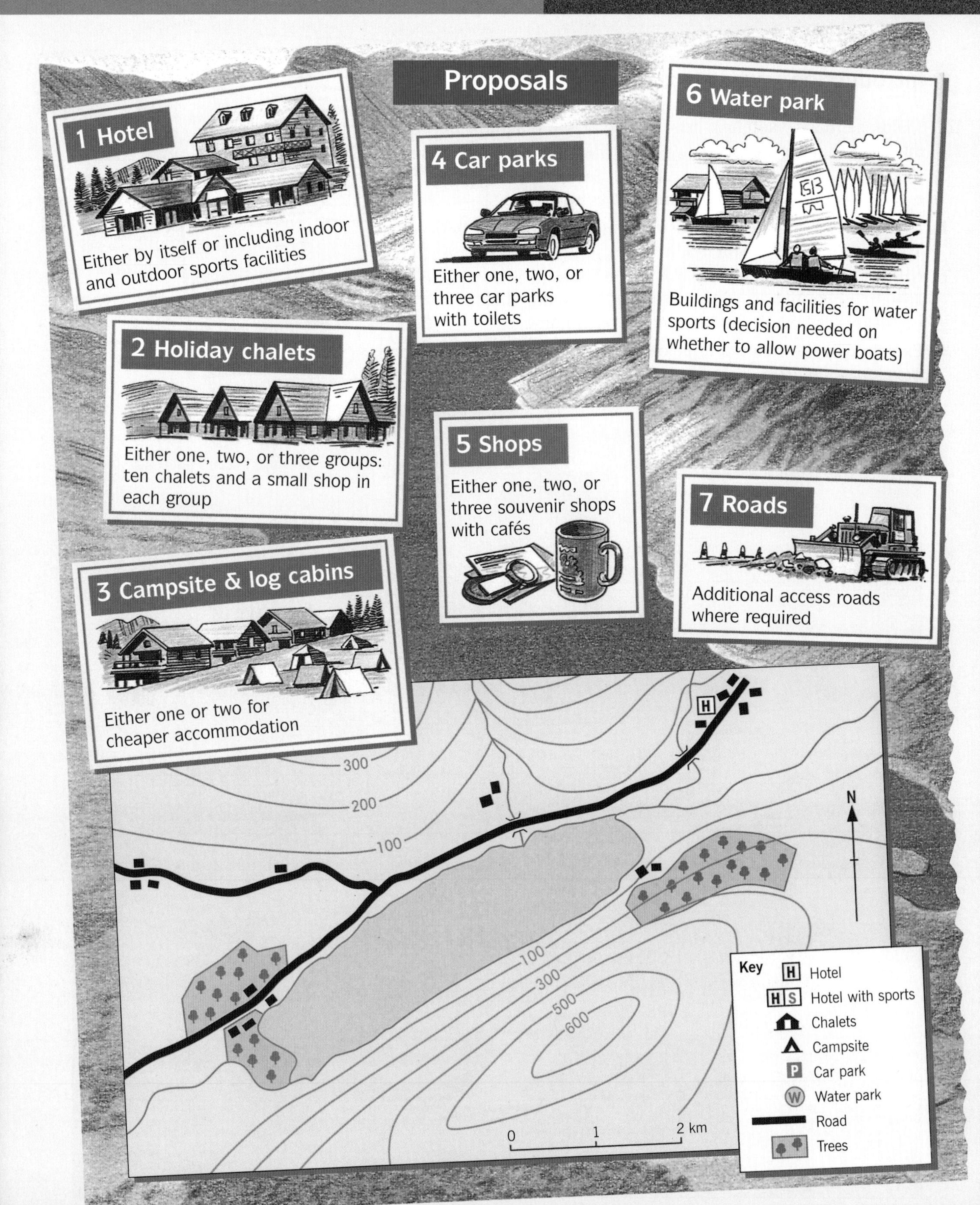

Proposals
1 Hotel
Either by itself or including indoor and outdoor sports facilities
4 Car parks
Either one, two, or three car parks with toilets
6 Water park
Buildings and facilities for water sports (decision needed on whether to allow power boats)
2 Holiday chalets
Either one, two, or three groups: ten chalets and a small shop in each group
5 Shops
Either one, two, or three souvenir shops with cafés
7 Roads
Additional access roads where required
3 Campsite & log cabins
Either one or two for cheaper accommodation
300
200
100
100
300
500
600
N
0
1
2 km
Key
H Hotel
HS Hotel with sports
Chalets
Campsite
P Car park
W Water park
Road
Trees

2 What will the development be like?

Look carefully at the proposals on the opposite page. They are designed to attract families with a broad range of interests and varying levels of wealth. Discuss with your partner or group what developments you wish to have, where they will go, and what they will look like. You can add other features of your choice if you wish.

Remember that your plans will need to be accepted by the National Park Planning Board. You should therefore consider very carefully the views of the park ranger on page 54.

- **a** On a copy of the map, mark your plans for the valley. Use the symbols in the key, and be as accurate as possible.
- **b** Describe and give reasons for your proposals. Mention any special points such as the positioning of features, materials to be used, and possible landscaping.

3 Conclusion

Now you must look carefully at your work and answer the enquiry question. Notice that it begins 'How ...'. That means you will need to both describe your findings, and explain them.

- **a** First you will need to make a summary of how the valley might change if it were developed for tourism.
- **b** Next you should list the good points and the bad points of the development. In doing this you should consider the local people, the tourists and the environment.
- **c** Finally you might like to decide whether the valley should be left in its present state or developed according to your suggestions. If it were developed, how could planning and management help to increase the good effects but reduce the bad ones? Page 52 will help you here.

4 Fashion and sport

How does fashion and sport affect our lives?

What is this unit about?

This unit looks at how companies, ideas and lifestyles are spreading more and more easily around the world. It uses fashion and sport as examples and shows how this globalisation, as it is called, can bring benefits but may also cause problems.

In this unit you will learn about:

- **globalisation and transnational corporations**
- **fashion and sport as worldwide industries**
- **links between countries at different stages of development**
- **the best location for a stadium**
- **the effects of a new stadium.**

A

B Athens 2004

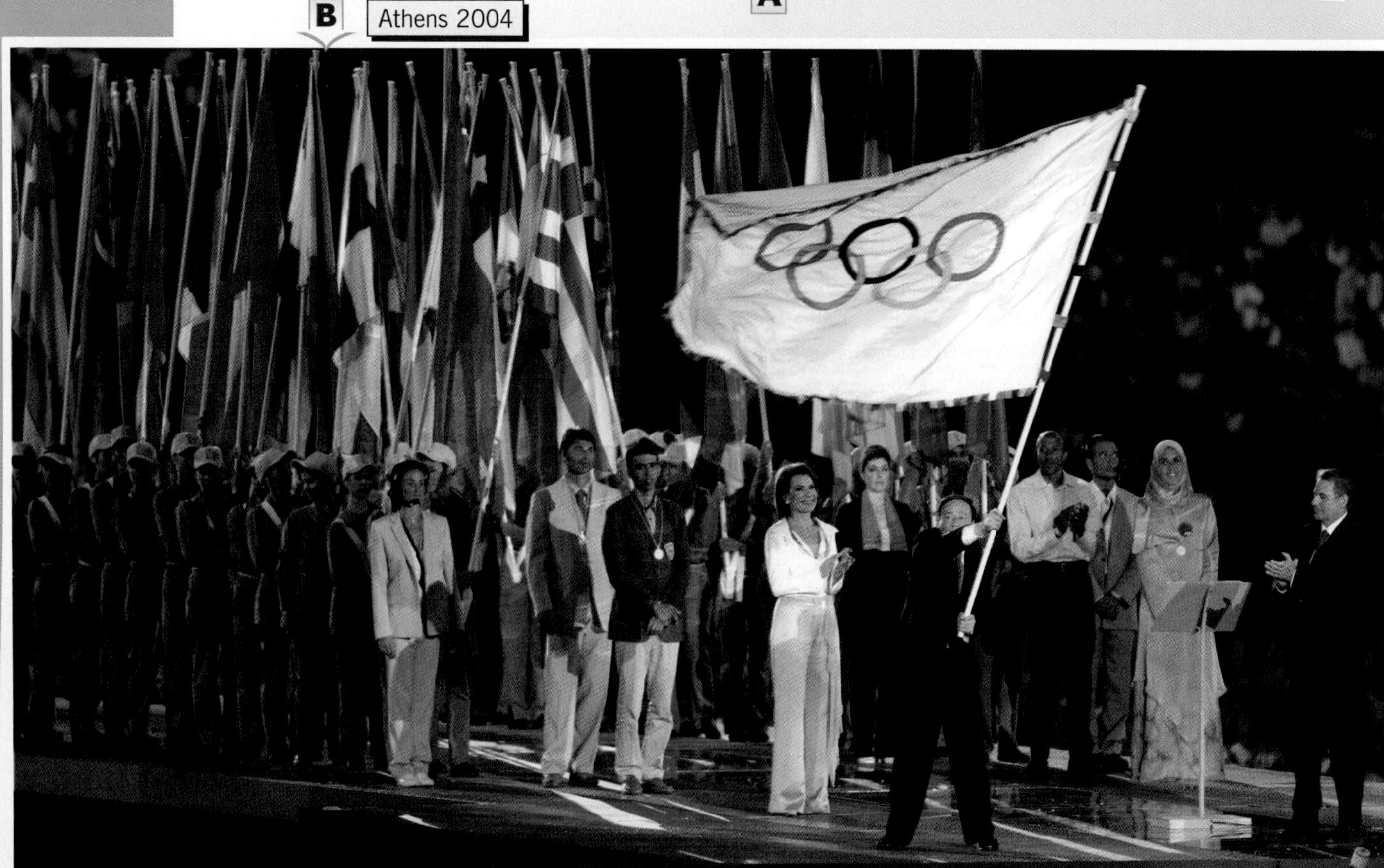

Why is learning about fashion and sport important?

Fashion and sport affect just about all of us in some way or another. Learning about them through geography can help us understand the links between people all around the world. It can also show us how they can affect people in many different ways. Some of the effects may be good but some can be bad.

This unit will also help you to:

- learn about two global industries
- think more carefully about what you buy and what you do and the effects that they may have
- appreciate that what happens in one country can affect the lives of people in other countries.

> - Are the goods in photos **A** and **C** fashion items, sports items or both? Who is likely to buy these goods? Give reasons for your answers.
> - Describe what is happening in photo **B**. In what ways can this be described as a 'world event'?
> - How can you tell that the stadium in photo **D** is modern?

C

D

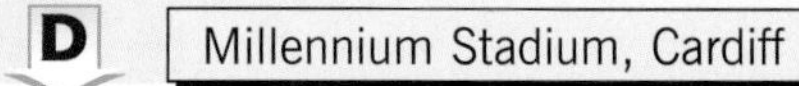
Millennium Stadium, Cardiff

What is globalisation?

You have probably heard the expression 'It's a small world'. People have been saying it for years but now it is true. Just check out the labels on your clothes – almost certainly they have been made in another part of the world. Turn on the computer and the internet will give you access to websites almost anywhere. Look at sport on television and you will see that it has a worldwide audience.

So why do we now live in a small world? The answer lies with improved travel and communications which have made links with other people and countries around the world so much quicker and easier. These links have increased at such a rapid rate that we now have a new word to describe it. The word is **globalisation**.

Globalisation means the way companies and ideas and lifestyles are spreading more and more easily around the world.

One important effect of globalisation is that it has made it much easier for goods and services produced in one place to be sold, used and seen elsewhere. Fashion and sport, the two topics in this unit, are good examples of this.

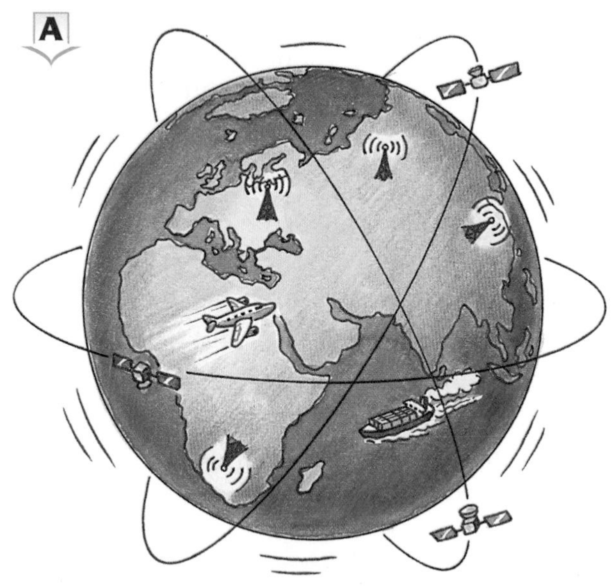

Look at the logos in drawing **B**. You probably recognise them as they belong to some of the biggest **brand names** in fashion. We buy their goods in our local shops, but they are actually made on the other side of the globe. In the same way, the sporting events shown in drawing **C** are based in several different countries but are immediately available around the world for everyone to watch.

To take advantage of globalisation, many large companies have become **transnational corporations** or **TNCs**. Transnational corporations have offices and factories all over the world. The headquarters are usually located in developed countries such as the USA or Japan. Smaller offices and factories tend to be in developing countries where labour is cheap and production costs are low. They have outlets to sell their products throughout the world.

Some TNCs are very large indeed. As tables **D** and **E** show, some are wealthier than many countries. Even Nike at the bottom of the list earned more money in a year than 90 of the world's 184 countries listed by the World Bank.

D Revenues for ten TNCs (in 2004)

TNC	Products	Revenue (US$ billions)
BP	oil/petrol	232
Exxon Mobil	oil/petrol	222
Ford	cars	164
Nestlé	food	65
Sony	electronics	63
Microsoft	software	34
Coca-Cola	food/drink	21
Gap	clothing	16
McDonalds	fast food	16
Nike	sports goods	11

E Money earned for ten countries (in 2004)

Country	Total wealth produced (GNP in US$ billions)
USA	11,667
Japan	4,623
UK	2,140
Italy	1,672
Brazil	604
Bangladesh	56
Kenya	16
Bolivia	9
Jamaica	8
Afghanistan	6

Source: World Bank

Activities

1 **a** What is meant by globalisation?

b Give three ways that globalisation affects you.

2 Each of the items in **F** helps the process of globalisation. Match each drawing with a statement from the list below.

- transports goods and people quickly
- improves links between people
- helps us see what is happening in the world
- transports goods cheaply
- provides information about the world
- improves world communications
- provides links and transfers information.

3 **a** Look at the brand labels on your clothes. Name the countries where they were made.

b Name at least four major sporting events not shown in **B**, that are on TV around the world.

4 **a** What is a transnational corporation?

b Suggest two reasons why TNCs have factories in developing countries.

F

Summary

Improvements in transport and communications have made it easier for companies, ideas and lifestyles to spread around the world. This is called globalisation.

How is the fashion industry changing?

The fashion industry has changed in recent years. It has become a growth industry and has taken advantage of globalisation by spreading its operations throughout the world.

So why has this happened? The main reason is that people living in richer countries have become increasingly well off and can afford to spend more money on clothes than in the past. They are able to buy clothes more often and can afford the more expensive, designer fashions that have become popular. Many are attracted by the **brand** or label as much as by the product itself.

It is these brand-name companies that have most increased their sales and are leading the way in seeking more profitable ways of manufacturing their products.

In the past, most clothing companies produced goods in their own factory. The traditional location of the factory was determined by the availability of transport and the nearness of raw materials, power sources, workers and markets for its goods.

This is now very different. Most large brand-name clothing companies have become transnational corporations that have offices and factories all over the world. They have found that going global reduces costs and increases profits. These companies include big names like Gap, Timberland, Reebok and Adidas. One of the largest and best known is Nike.

Nike is a typical TNC. It has its main offices in the USA and its production lines in developing countries where labour is cheap and costs are low. As the designs and styles of trainers and sportswear constantly change, it is also cheaper to employ manual workers, who are readily available in these places, than machines. This is because it is easier, faster and cheaper to get employees to adapt to the new designs than to change or buy new machinery.

Diagram **A** and map **B** show how Nike operates on a global scale.

Headquarters – Oregon, USA
Nike are based in Portland, Oregon where most of the product design, marketing and administration is done. No goods are produced here or anywhere else in the USA.

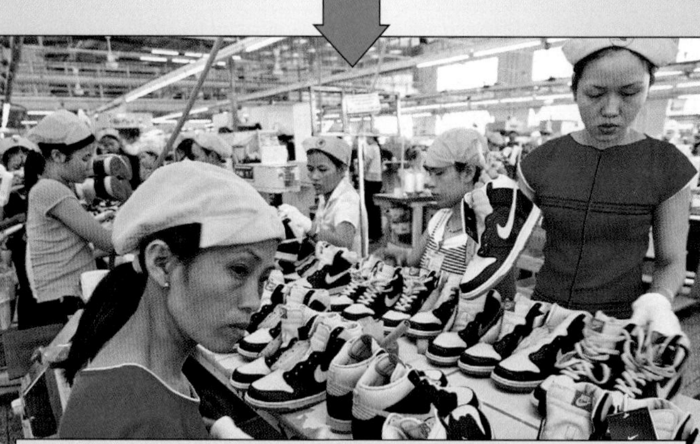

Production – mainly in South-east Asia
Nike has factories in 40 different countries and employs over 500,000 people mostly on low pay. Earnings are as low as £2.50 a day. Trainers which can be made for £1 sell in the UK for over £90.

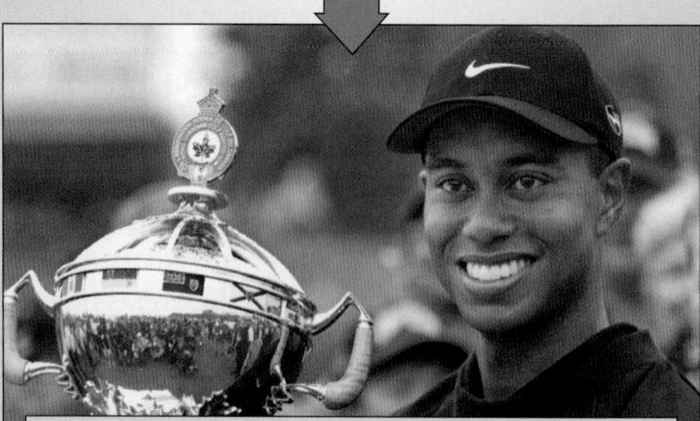

Sales and advertising – worldwide
Nike supplies goods to over 47,000 shops in 140 different countries. It spends up to a billion pounds a year on advertising which includes paying superstars like Tiger Woods to wear Nike products.

Nike, a typical transnational corporation. **A**

B The world of Nike

1 Began in Oregon in the 1960s as 'Blue Ribbon Sports'. It distributed trainers made in Japan before changing its name in 1972 to Nike. It now designs and markets sports shoes and sportswear. It has no production factory in the USA.

2 In the 1960s, training shoes were made in Japan where labour was cheap.

3 By the late 1970s Japanese wages had risen so Nike moved its production to South Korea and Taiwan.

4 In time, wages in South Korea and Taiwan rose. So, in the late 1980s, Nike moved to Thailand and Indonesia.

5 Since the 1990s, over 50% of total production has come from China and Vietnam.

6 Most Nike products are sold in North America and Europe.

Portland, Oregon

USA

Key
- ■ Nike headquarters

Countries with Nike:
- ● branch offices
- production lines

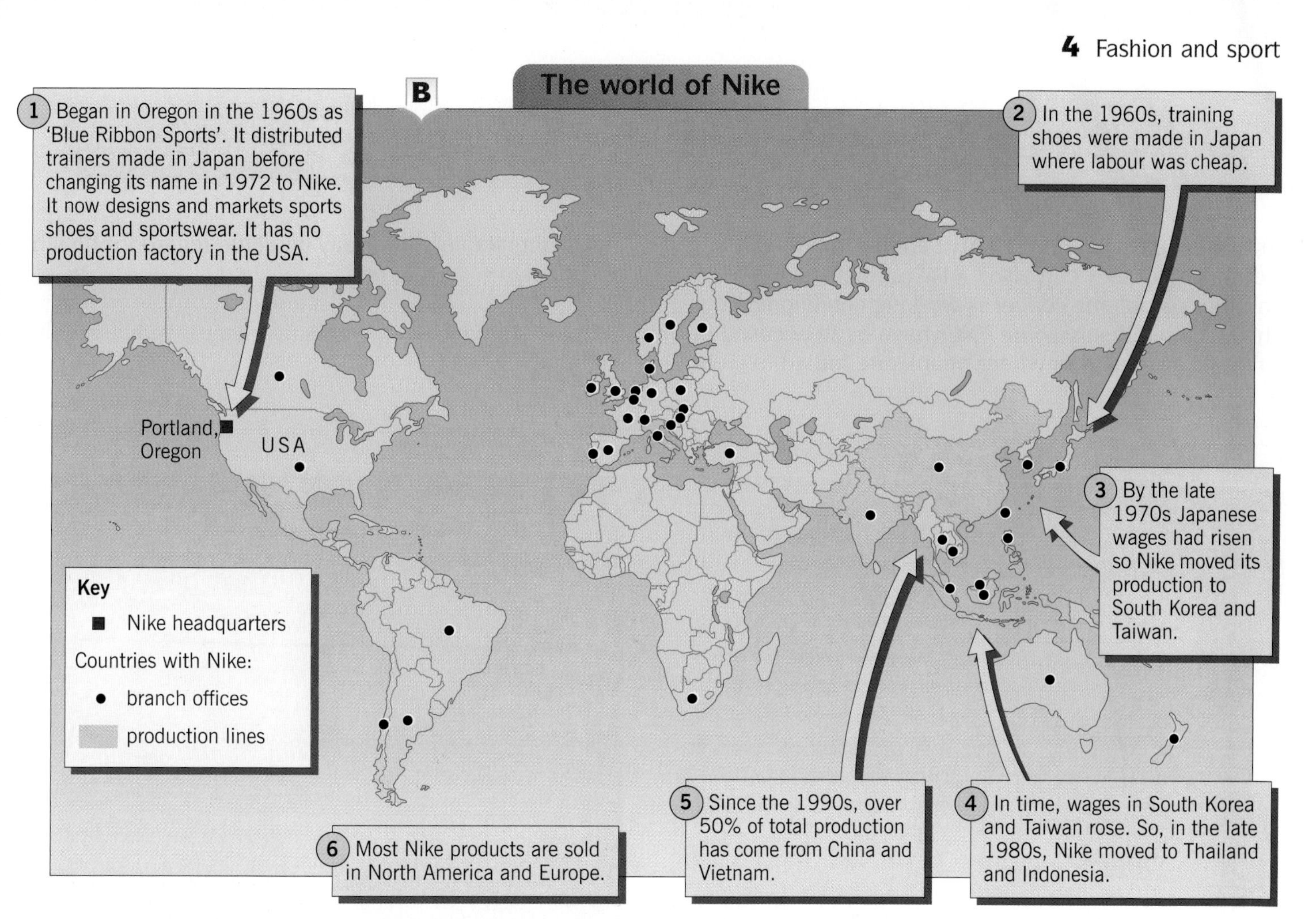

1 Describe two changes in the fashion industry. Suggest reasons for the changes.

2 **a** Make a larger copy of diagram **C** and complete it by adding five location factors.

b Make a larger copy of diagram **D** and complete it by putting the following in the correct places.

- Mainly rich countries
- Low paid
- Worldwide
- Design and marketing
- Developing country
- Developed country

3 Answer the following by using map **B** and the world map on the back cover.

a Name six countries where Nike has shops.

b Name six countries where Nike goods are made.

c Name four countries in the developing world where Nike has branches and where goods are made.

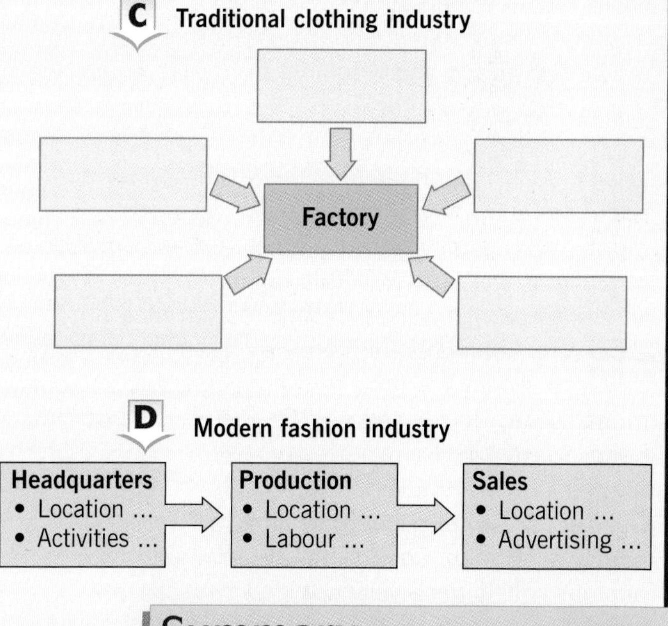

Summary

Increased wealth and the popularity of brand image have helped fashion become a growth industry. Many large companies have gone global to reduce costs and increase profits.

How do transnationals affect poorer countries?

Transnationals can bring many benefits to poorer countries but they can also cause problems. One of the main problems concerns working conditions. In the fashion industry some TNCs have been accused of creating **sweatshops** where people are forced to work long hours for very poor pay and little regard for health and safety.

Merpati, shown below, works in a typical sweatshop in Jakarta.

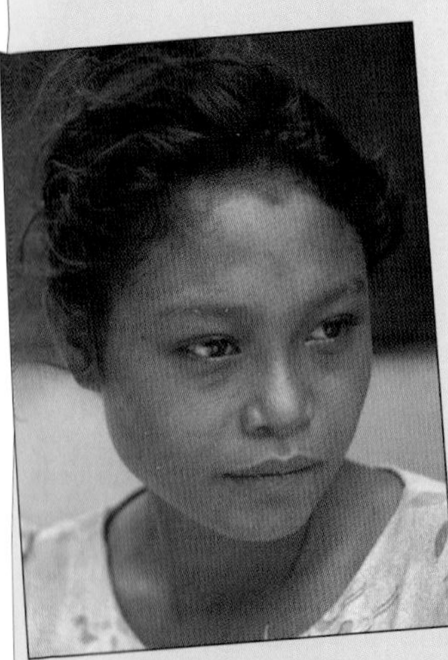

Hello,
My name is Merpati and I live in Jakarta, Indonesia. I am 16 years old and work in a factory making clothes for a well-known fashion label. Until last year I lived with my family in a small rural village 125 km from here. There was no work there so I had to leave.

The clothing company runs my life. I live in a dormitory behind the factory and share a small room with 11 other women. We sleep in bunk beds and have one small cupboard each. There is little privacy or space and washing and toilet amenities are shared with 50 other people.

I don't spend much time in the dorm as I always seem to be working. I only get two days off a month and usually work 14 hours a day from 7.30 in the morning until 9.30 at night. At peak times we are forced to work overtime, sometimes until 2.30 a.m. I earn about £3 a day but the company deducts money for food and accommodation so my monthly pay is less than £70. I send most of this home to my family.

Our factory employs several hundred workers, nearly all women. Conditions are cramped and unpleasant. It gets very hot and the jobs are repetitive and boring. I worry about my health and am always tired. The company is very strict and has many rules. For example, we get fined for getting a drink, talking, being late or refusing to work overtime.

A

B

I would like to go home to my village but I can't. I need to earn money and must have a place to live. My biggest worry is that the factory will close and then I would be out of work.

Not all the clothes factories in developing countries are as bad as the sweatshop where Merpati works. Indeed, as photo **C** shows, many are now more modern than some factories in the UK.

One of the reasons for this improvement is that companies like Nike, Adidas, Gap and others now have rules to control working conditions in factories producing their goods. These rules are a response to criticism that they have been **exploiting** workers in poorer countries in order to increase their profits.

Whilst conditions are now much improved in some places, there are many workers like Merpati who are still being exploited.

C

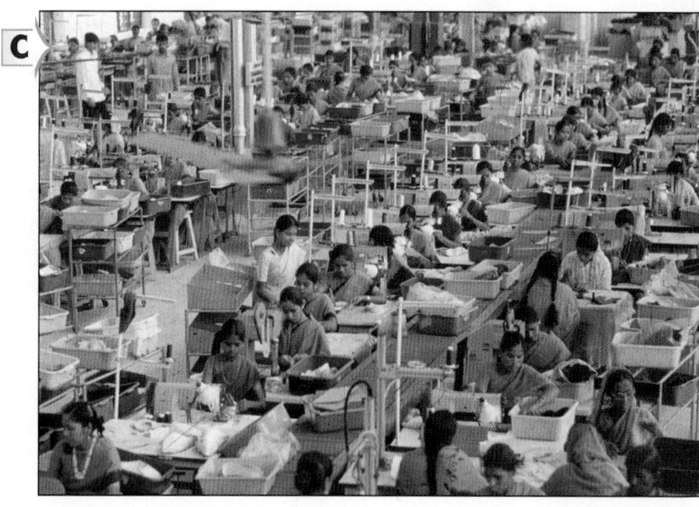

D

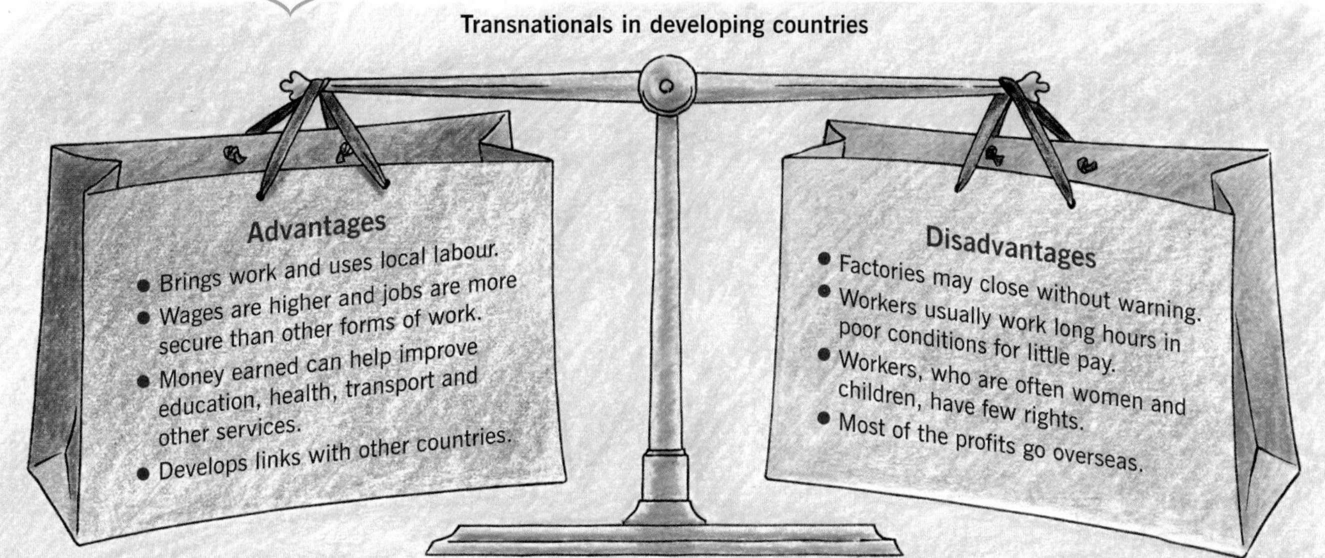

Activities

1. **a** What is a sweatshop?

   b People working in sweatshops are said to be exploited. What does 'exploit' mean?

2. **a** Briefly describe what life is like for Merpati. Use these headings:
   - Living conditions
   - Working hours and pay
   - Working conditions.

   b Which part of her life would you like least?

3. Why do you think Nike and other companies have begun to do something about sweatshop conditions? Give at least two reasons.

4. Complete the speech bubbles in **E** to give two different views of transnational companies in developing countries.

E

I'm in favour of transnationals coming to our country because ... and ... .

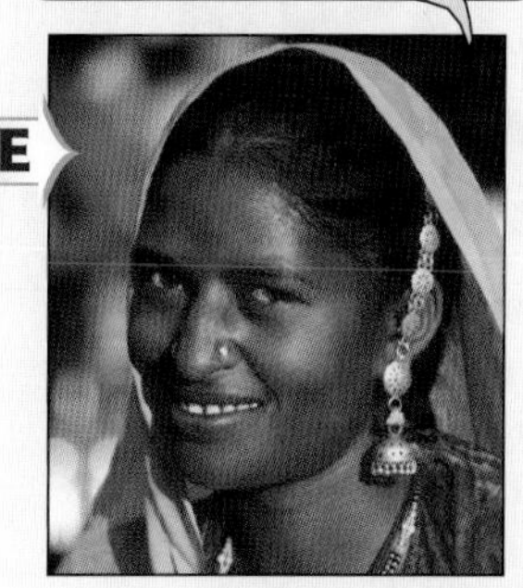

I'm against transnationals coming to our country because ... and ... .

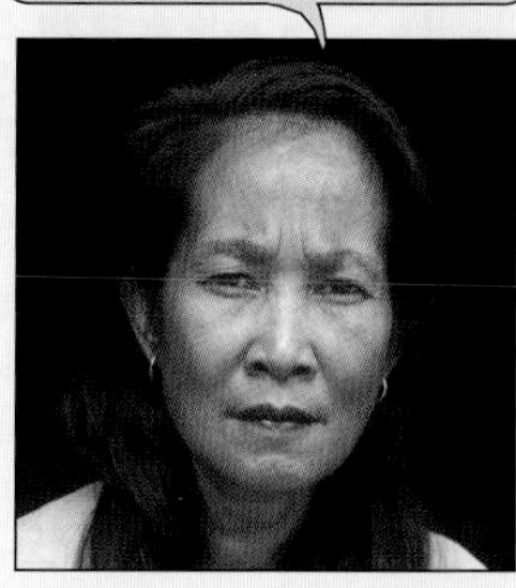

Summary

Transnational corporations can bring benefits to developing countries but can also cause problems. Care must be taken to ensure that transnationals do not exploit workers in poorer countries.

What has happened to the clothing industry?

During the nineteenth century Britain was the world's main producer of clothing and textiles. Thousands of people were employed in mills making cotton goods in Lancashire and woollen goods in Yorkshire. Later in the twentieth century many people came from places like India, Pakistan and Bangladesh to work in these mills. Many settled here, had children and became British citizens.

The situation has now changed. As we have seen, developing countries, particularly in South-east Asia, now have their own clothing and textile industries and can produce goods much more cheaply than we can in Britain. This has caused problems for the clothing industry, some of which are shown in the news reports from 2002.

A

Shock news from M&S

The largest clothing retailer in the UK, Marks & Spencer, sent shockwaves through the industry when it announced that it planned to buy the majority of its products from overseas.

M&S has always had a 'Buy British' policy with 70% of its clothing made in the UK. The company aims to cut costs by purchasing in cheaper markets abroad. Marks & Spencer say the move should help it reduce prices, increase profits and retain staff in its UK shops.

Clothing industry hit by more factory closures

The textile manufacturer Dewhirst is closing two factories with the loss of over 1,000 jobs. The closure is part of an ongoing programme to cut costs and move production overseas to places like China, Indonesia and Malaysia where most of the firm's manufacturing is now done.

The GMB Union blamed the job losses on the decision by Marks & Spencer to move its source of product manufacturing abroad. Dewhirst supplies 90% of its production to Marks & Spencer.

Job losses blamed on labour costs

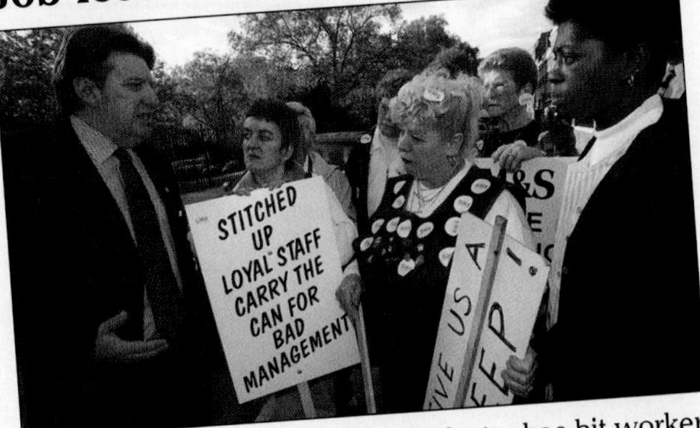

The decline of the UK clothing industry has hit workers throughout the country. Several thousand jobs have been lost as manufacturers throughout the country have either closed down or moved their production lines abroad where labour costs are lower.

One long-time machinist working in the north-east said, 'It's crazy, we are on a minimum wage of £4.50 an hour but the company's new workforce in China will be paid less than 20p an hour. How can we compete?'

In 2005, people in Britain spent £31 billion on clothes. That is about £520 for every man, woman and child. Not so long ago, most of those clothes were made in the UK. Nowadays almost three-quarters are made in developing countries.

The effects of these changes on the UK clothing industry have been considerable. Many companies have closed down whilst others have had to relocate their factories overseas. Thousands of workers have lost their jobs. The changes have also had an effect on UK trade figures. As graph **C** shows, imports have a much higher value than exports and this difference is expected to increase even further in the future.

Efforts have been made to protect the UK clothing industry. In 2005, the EU and China agreed to put temporary limits on textile imports into Europe. It was hoped that this would reduce the problem for a few years and give manufacturers time to adapt to the new conditions resulting from globalisation.

B Clothes shopping in Marks & Spencer

C UK exports and imports of clothing

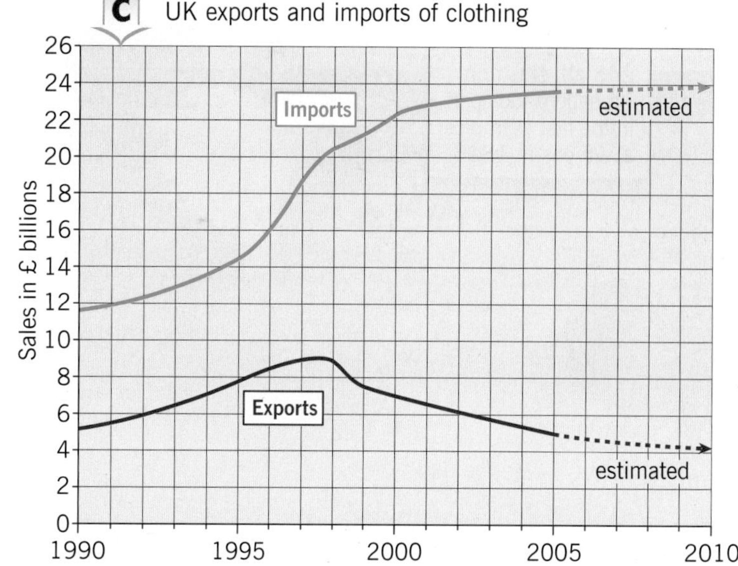

Activities

1 **a** Why did M&S decide to buy overseas?

b Give three advantages of the move.

c Why did the news shock the clothing industry?

2 Look at the placards in drawing **D**. Which of the following would support the views of the protestors?

- Clothes shoppers
- M&S management
- A worker at Dewhirst
- A worker in a sweatshop in China

Give reasons for your answers.

3 Look at graph **C**.

a Copy and complete table **E**.

b Describe the trends in imports and exports.

c When was the greatest change in exports and when was the greatest change in imports?

Suggest reasons for each of these changes.

D

E

Year	1990	2000	2005	2010 (est.)
Exports				
Imports				

Summary

Cheaper imports from developing countries have led to factory closures and a loss of jobs in the UK clothing industry.

How worldwide is sport?

The Olympic Games is an international sporting event which takes place every four years. The first modern games were held in Athens in 1896 when just 15 countries took part. Since then the games have grown in size and popularity. Almost every country now takes part and through television, events can be seen live, virtually anywhere in the world.

After Beijing in 2008, the games will of course be heading for London. They were last held here in 1948 and will return again in 2012. Map **A** shows the Olympic venues between those dates.

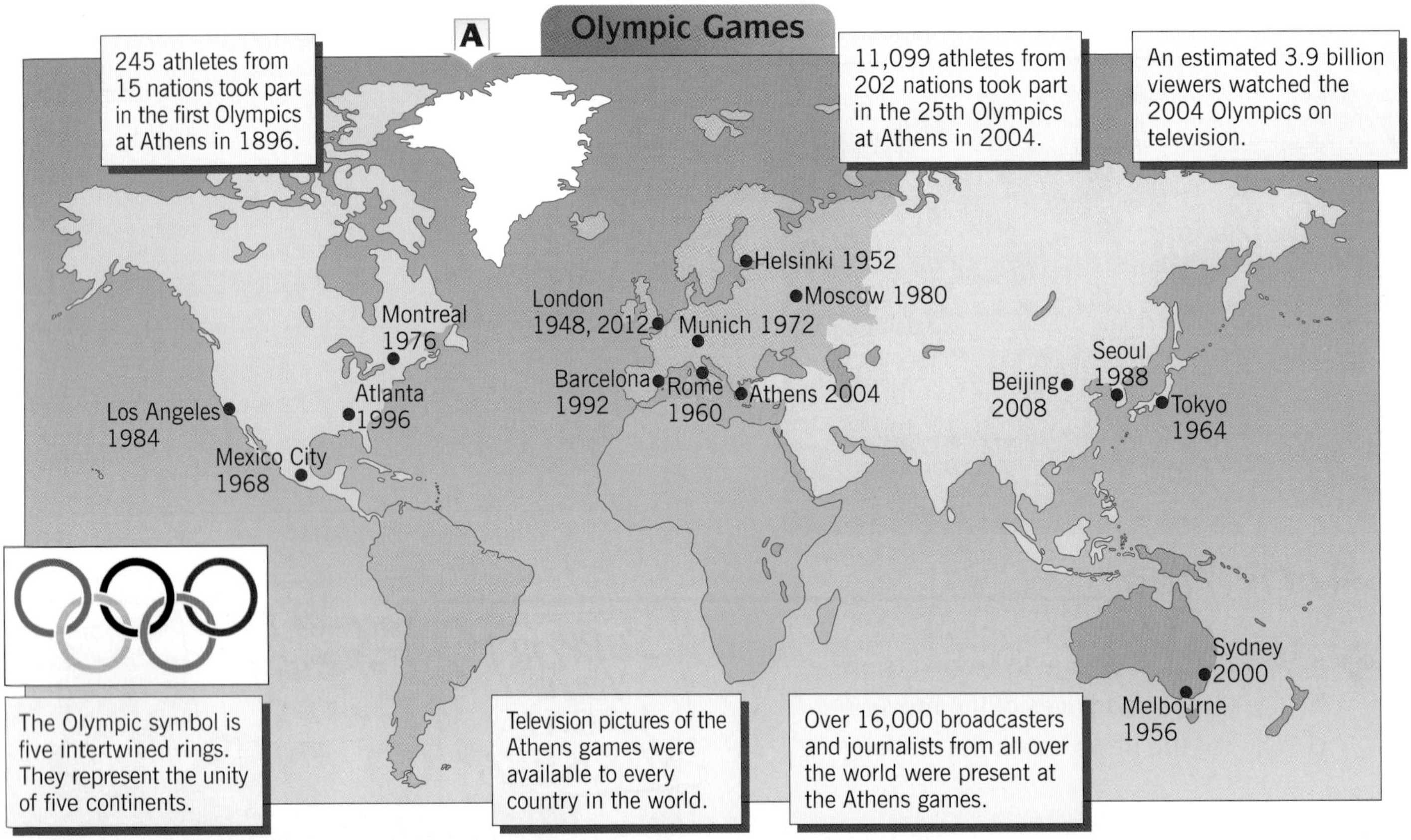

A Olympic Games

245 athletes from 15 nations took part in the first Olympics at Athens in 1896.

11,099 athletes from 202 nations took part in the 25th Olympics at Athens in 2004.

An estimated 3.9 billion viewers watched the 2004 Olympics on television.

Helsinki 1952
Moscow 1980
London 1948, 2012
Munich 1972
Montreal 1976
Atlanta 1996
Los Angeles 1984
Mexico City 1968
Barcelona 1992
Rome 1960
Athens 2004
Beijing 2008
Seoul 1988
Tokyo 1964
Sydney 2000
Melbourne 1956

The Olympic symbol is five intertwined rings. They represent the unity of five continents.

Television pictures of the Athens games were available to every country in the world.

Over 16,000 broadcasters and journalists from all over the world were present at the Athens games.

Activities

1 **a** Make a copy of table **C**.

b List the Olympic venues from 1948 to 2012. For each one give the country and continent. The world map on the back cover will help you.

C Olympic venues, 1948 to 2012

Year	City	Country	Continent
2012	London	UK	Europe
2008			

D

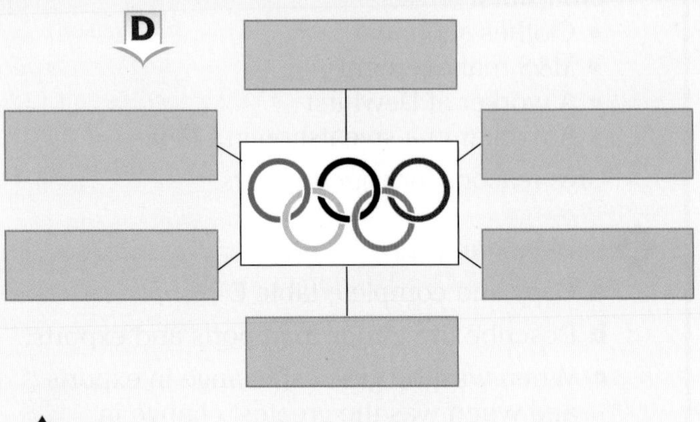

2 The Olympics is often described as the world's greatest global event. Suggest reasons for this. Put your answers in a star diagram like **D** above.

The growth of the Olympics has largely been due to globalisation. Improved travel has made it easier to attend the event whilst better communications have given the games a worldwide audience on television.

Globalisation has had a similar effect on football. The Premiership is watched by over a billion people worldwide and matches are shown on a regular basis in 150 countries. This has brought great wealth to the sport with total club earnings of over £1.3 billion a year. Much of the money is from television fees, selling football shirts and through sponsorship by companies like Nike.

The Premiership has also become global in terms of its players. Over 280 foreign-born players now compete in the league. They come from 62 different countries and make up some 45 per cent of the total number of players in first team squads.

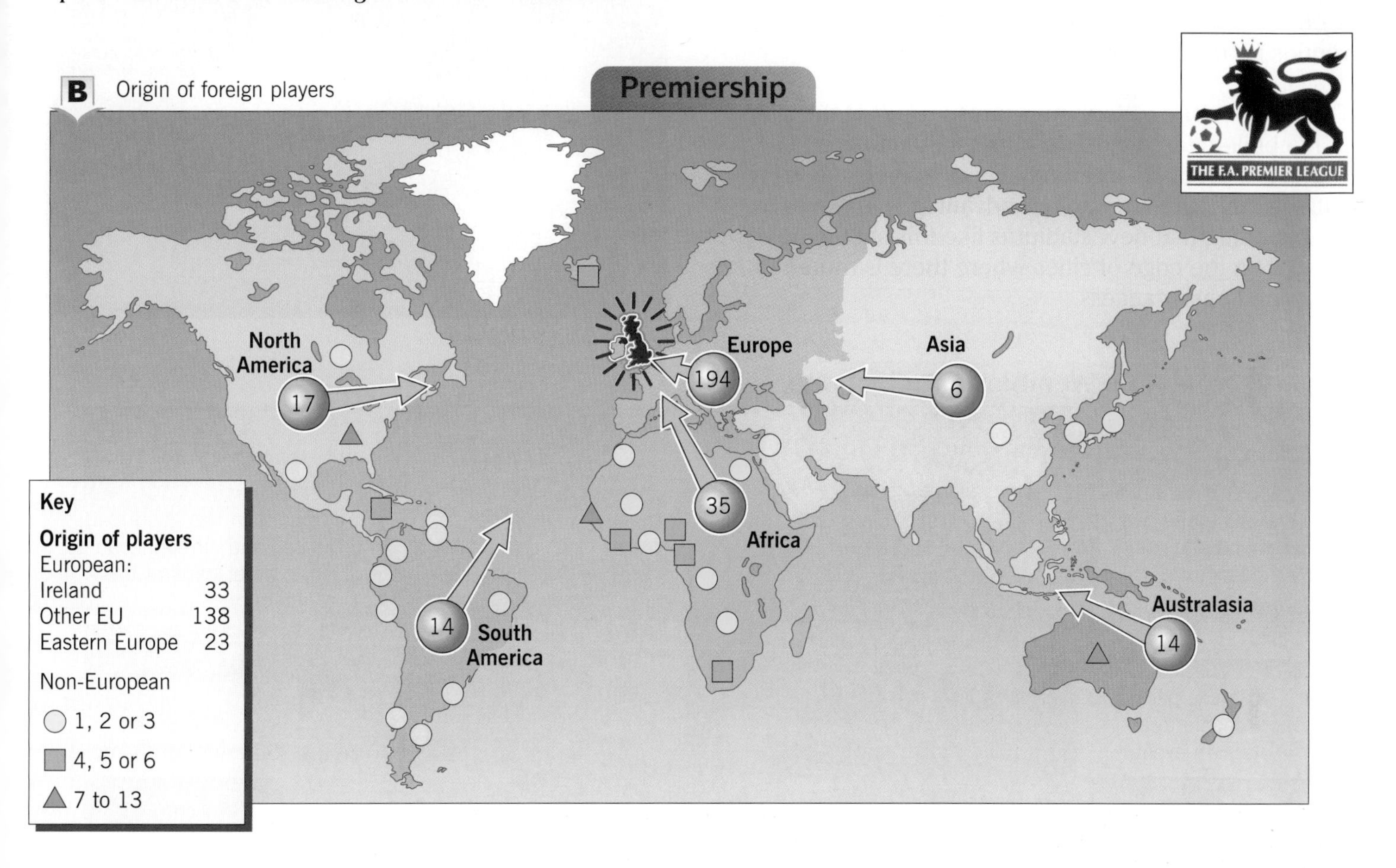

B Origin of foreign players

Premiership

THE F.A. PREMIER LEAGUE

North America 17
Europe 194
Asia 6
Africa 35
South America 14
Australasia 14

Key

Origin of players

European:

Ireland	33
Other EU	138
Eastern Europe	23

Non-European

○ 1, 2 or 3

□ 4, 5 or 6

△ 7 to 13

a Make a copy of table **E**.

b Complete the table by choosing two countries from each continent apart from Europe. The world map on the back cover will help you.

E Origin of foreign players

Continent	Country	Number of players
Asia		
Asia		
Africa		

4 Many Premiership teams visit the USA and Far East on pre-season tours. Suggest why they do this. Try to give at least three reasons.

5 Imagine that you are a young footballer just beginning your career. What would be the advantages and disadvantages of having foreign players at your club?

Summary

Improvements in transport and communications have helped sport become a global activity that is watched throughout the world.

Where is the best site for a stadium?

Photo **B** shows the location of the new Wembley Stadium. It opened in 2007 on the site of old Wembley. It cost over £757 million to build and has seating for 90,000 spectators. The stadium will be the home of English football and will host a variety of sporting events and live concerts. It will also be one of the venues for the London 2012 Olympics.

The location of the stadium is unusual in that the site is in a built-up area close to houses and industry. Traffic is a problem and road access to the stadium is difficult. Although the site has advantages, many people argue that new stadiums like this should be located on the edge of cities where there is more space and easier access.

A Wembley Stadium

B The new Wembley Stadium site

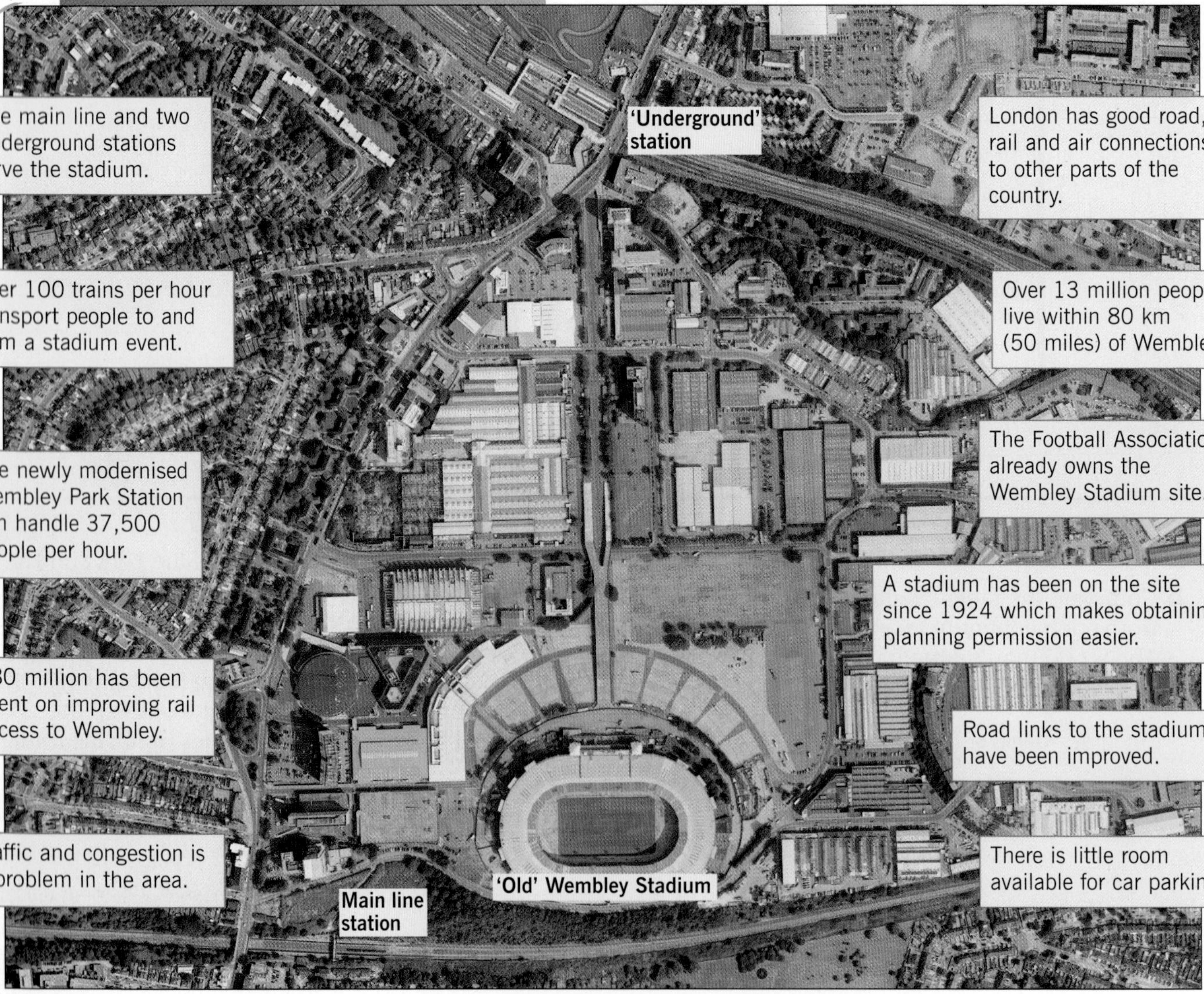

One main line and two Underground stations serve the stadium.

'Underground' station

London has good road, rail and air connections to other parts of the country.

Over 100 trains per hour transport people to and from a stadium event.

Over 13 million people live within 80 km (50 miles) of Wembley.

The newly modernised Wembley Park Station can handle 37,500 people per hour.

The Football Association already owns the Wembley Stadium site.

A stadium has been on the site since 1924 which makes obtaining planning permission easier.

£80 million has been spent on improving rail access to Wembley.

Road links to the stadium have been improved.

Traffic and congestion is a problem in the area.

'Old' Wembley Stadium

Main line station

There is little room available for car parking.

One of the locations suggested as an alternative to Wembley was a site on the outskirts of Birmingham. This site had the advantage of good motorway access, extensive car parking facilities, cheap land values and a pleasant environment. There were, however, disadvantages to the site. One of these was accessibility due to a lack of public transport.

A report published at the time estimated that 90,000 fans attending the new Wembley would be able to leave the stadium largely by public transport within one hour. For the Birmingham option, heavily dependent on cars and coaches, it would take four times as long and result in massive congestion on the M40 and M6 motorways.

Maps **C** and **D** compare the two locations.

C Birmingham

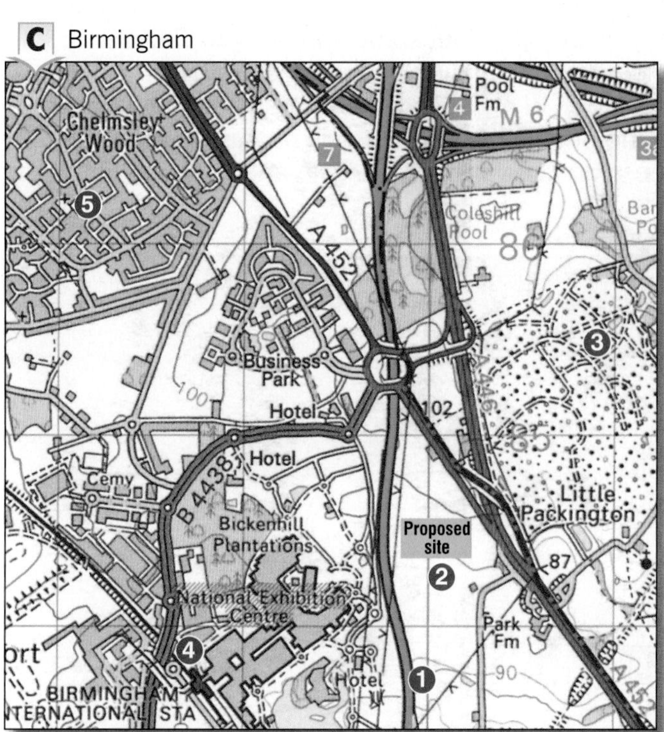

D Wembley

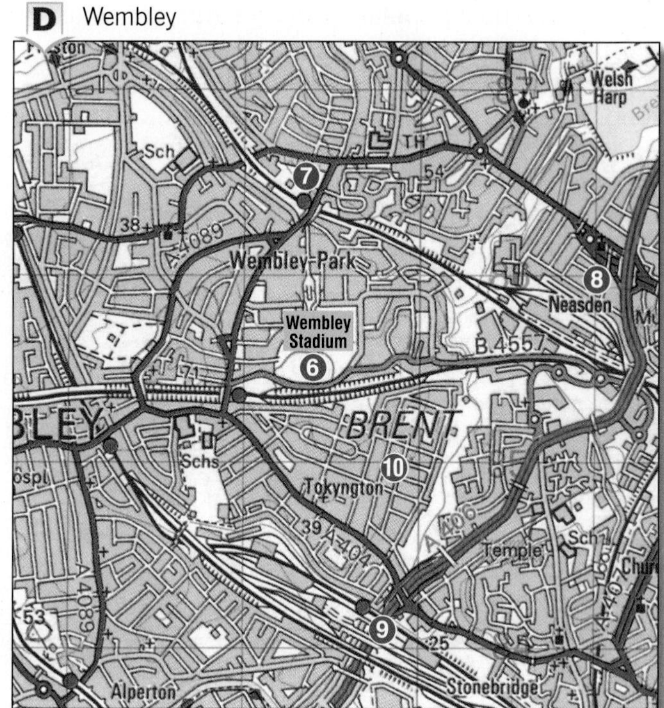

1. Match each of the features on drawing **E** with a number from maps **C** and **D**.

2. Make a larger copy of drawing **F** and complete it to show the advantages and disadvantages of the site of the new Wembley Stadium.

F

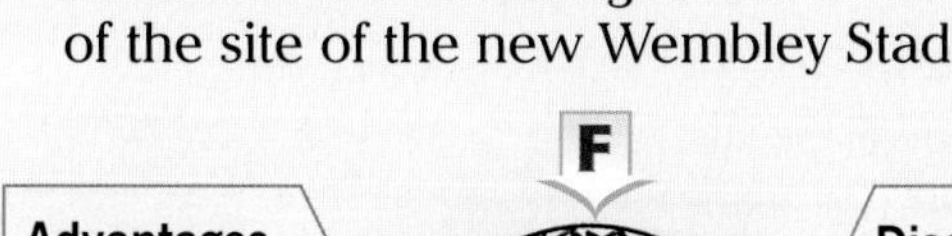

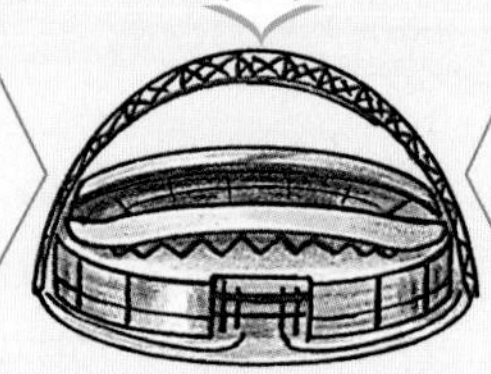

Wembley Stadium

3. What are the advantages and disadvantages of locating a stadium on the outskirts of a city such as Birmingham?

E

Location factors
- Motorway access
- Site of 'old' Wembley
- Large population nearby
- Birmingham nearby
- Road and rail links to rest of country
- Underground rail access
- Pleasant environment
- Road links improved
- Rail access
- Land available

Summary

A stadium should be located where there is cheap flat land, good transport facilities and a nearby large town. Wembley has good public transport but suffers from difficult access by road.

Stadiums: the good news ... and the bad news

Sport is big business and operates on a global scale. Huge **transnational corporations** such as Nike, Sky Television and Coca-Cola invest hundreds of millions of pounds in sport every year.

Football, the most popular sport on earth, generates an estimated £147 billion a year and provides employment for thousands of people around the world. The finals of the last World Cup, which lasted four weeks, attracted a total television audience of over 33 billion people worldwide, making it the most viewed event in television history.

It is not surprising, therefore, that sport affects just about all of us in some way or another. Most of the effects are good but some can be bad. Take for example the building of a new stadium or the extension of an existing one. This can bring benefits to some people but cause problems for others. Some of these are shown in drawings **A** and **C**.

A Benefits

- A new stadium and a successful team brings prestige to the city.
- It creates additional full-time jobs in and around the stadium.
- It provides part-time jobs on match days and at other events.
- It creates hundreds of other jobs that are sport dependent.
- It helps local businesses like public houses, cafés and restaurants.
- It brings money into the city and increases wealth.
- It provides a social occasion when friends can meet for a match.

Activities

1 Look at drawing **D** which shows some of the jobs created by football.

a Copy table **E** and sort the jobs into the correct columns.

b Add at least four more jobs to each column.

c Apart from being a player, which job would you like to do? Give reasons for your answer.

E

Jobs at the club		Jobs outside the club	
Full-time	Part-time	Media	Others

D

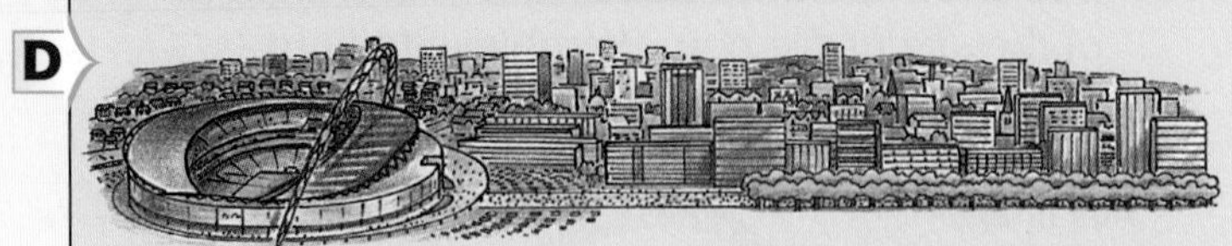

- Football coach
- Football shirt maker
- Programme seller
- Footballer's agent
- Football strip designer
- Match-day steward
- Radio commentator
- Physiotherapist
- Football writer
- Restaurant waitress
- Groundsman
- Receptionist

Whilst the building of a sports stadium will mainly affect just the local community, large events such as the football World Cup and the Olympic Games can have much wider effects.

It is hoped, for example, that the London 2012 Olympic Games will help **regenerate** some of the derelict areas of east London and improve the quality of life for people living there. It is also hoped that London itself and the UK in general will benefit from investment in the games and from the publicity and increased tourism that the event will generate. Some people are concerned, however, that the Olympics may cause problems as well as bring benefits.

B Crowds celebrating London's successful bid for the 2012 Olympics

C Problems

- A large stadium can be an eyesore and spoil the environment.
- It can take up land that could be used for other purposes.
- Football crowds are noisy, leave litter and can be difficult to control.
- A minority of spectators may drink too much or cause vandalism.
- On match days there may be serious traffic congestion on local roads.
- Thousands of supporters can fill nearby shopping centres.
- Cars may be parked outside local residents' houses, blocking access.

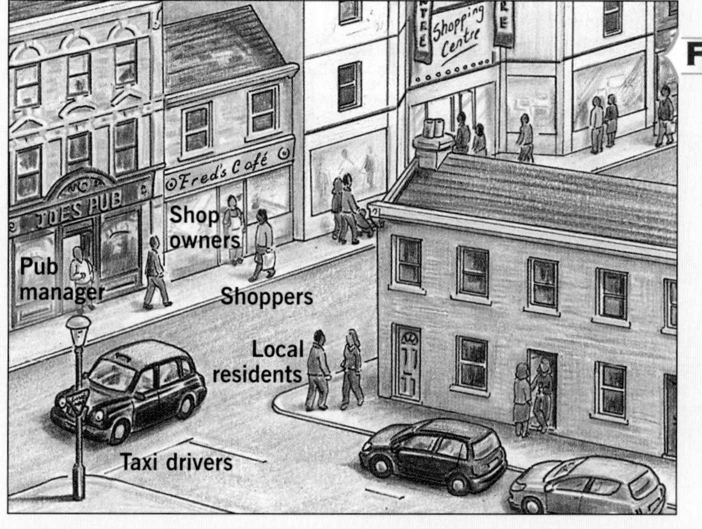

F

2. Look at the people shown in drawing **F**. Who do you think will be in favour of a large sports stadium nearby, and who will be against it? Give reasons for your answers.

3. Look at photo **B**. Why do you think the people are so pleased about what has happened?

Summary

Sport affects just about all of us in some way or another. New stadiums make watching sport more enjoyable and help clubs make more money. They also bring other benefits but can cause problems.

The Olympics enquiry

Most of the London Olympics will be located in the East End of the city close to Stratford. This is one of the most deprived areas of London and one that is in real need of **regeneration**. The area is generally run-down with poor housing, high unemployment and a lack of basic services. Derelict buildings and wasteland are also a problem.

One of the main aims of the London Olympics is to help regenerate the area and improve the quality of life of people living there. Your task in this enquiry is to make a report on how this can be done.

How can the Olympics benefit London?

1 Introduction

a Look at the information on the back cover. Draw a simple sketch map to show the Olympic Park. Mark and label the main features in 2006 and the developments planned for 2012.

b Describe the problems of the area in 2006.

2 Main part

a Look at drawing **C**. Draw a star diagram to show how the Olympics can help regenerate the area.

b Look at drawing **B** and arrange the comments in a diamond pattern as shown in diagram **A**.

3 Conclusion

Now look carefully at your work and answer the enquiry question. You could use these headings:

- Reason for regeneration
- Benefits of the Olympics
- Main difficulties
- Expected outcomes

Pages 82 and 83 will help you with this.

A

1 Most important benefit

2 Next most important benefits

3 Biggest problem

4 Next biggest problems

5 Remaining comments

B London Olympics: some benefits and problems

C

London Olympics 2012

Stadiums There will be several new arenas. The main one will seat 80,000 and will host the main ceremonies and athletics events.

Transport £21 billion will be spent on improving London's transport and creating Stratford International rail terminal.

City development A new centre with shops, schools, health centres and 5,000 homes will be built on previously derelict land.

Jobs Many new jobs will be created. On site there will be 7,000 in construction, 12,000 permanent and 43,000 temporary during the games.

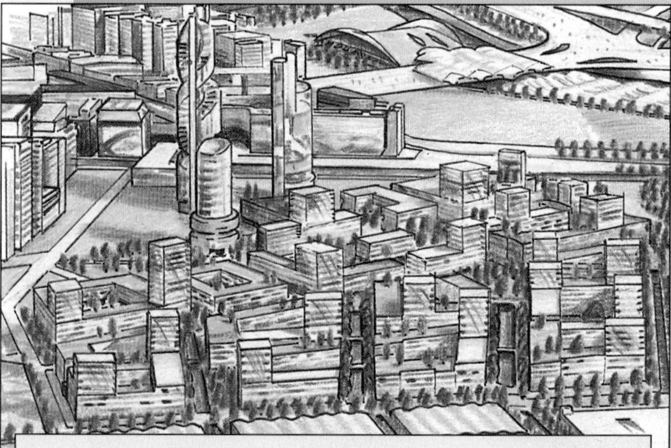

Olympic village This will house 17,000 athletes during the games. Afterwards it will be converted mainly to low-cost housing for local people.

Parkland After the games the Lower Lea Valley will be transformed into the largest urban park created in Europe for more than 150 years.

5 Italy, an EU country

What is Italy like?

What is this unit about?

This unit is about Italy, a country with a long and interesting history and its own customs and way of life. The unit is in two parts. The first looks at Italy's main features and the second at regional differences between the North and the South.

In this unit you will learn about:

- **the European Union (EU)**
- **Italy's main physical and human features**
- **the country's main regions**
- **differences between the North and the South**
- **how developed Italy is.**

Why is learning about Italy important?

This unit on Italy will give you an interest in and knowledge of people and places that are different from those found in the UK. It will also help you learn more about a country that plays an important role in Europe as well as being one that you may well like to visit in the future.

This unit will also help you to:

- broaden your knowledge of the world
- learn about different landscapes and climate
- understand ways of life that are different from your own
- recognise differences within a country
- develop an interest in other countries.

B Naples and Mount Vesuvius

B Venice

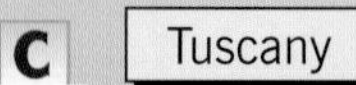

C Tuscany

- Compared with where you live, how different is:
  - the city and landscape in photo **A**
  - the buildings and activities in photo **B**
  - the countryside in photo **C**?
- Which of the places:
  - would you most like to visit
  - is most different from where you live?

  Give reasons for your answers.

What is the European Union?

The European Union, or EU in short, is a group of countries trying to work together. It began when six countries joined together to try to build up their industries and improve their economies following the Second World War. Since then it has expanded in two ways.

1 The number of member countries has increased to twenty-five. More countries, mainly in Eastern Europe, are likely to apply to join in the next few years.

2 Its activities have grown from trade and industry to include finance, tourism and care of the environment. Other activities are shown in diagram **C**.

A

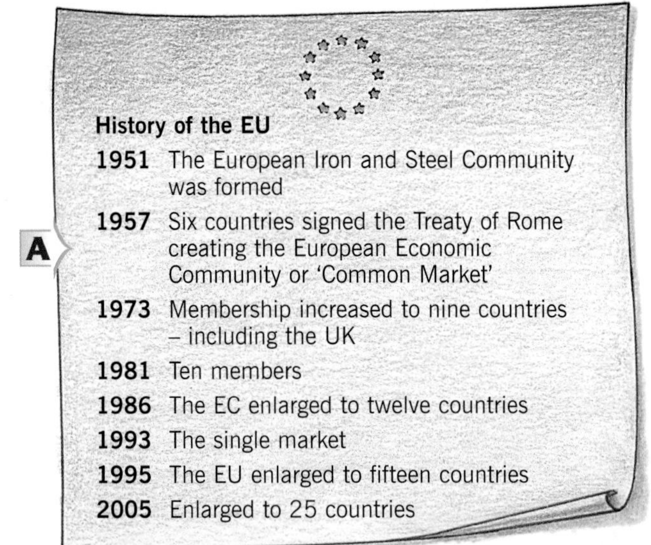

History of the EU

1951	The European Iron and Steel Community was formed
1957	Six countries signed the Treaty of Rome creating the European Economic Community or 'Common Market'
1973	Membership increased to nine countries – including the UK
1981	Ten members
1986	The EC enlarged to twelve countries
1993	The single market
1995	The EU enlarged to fifteen countries
2005	Enlarged to 25 countries

B Members of the EU in 2006

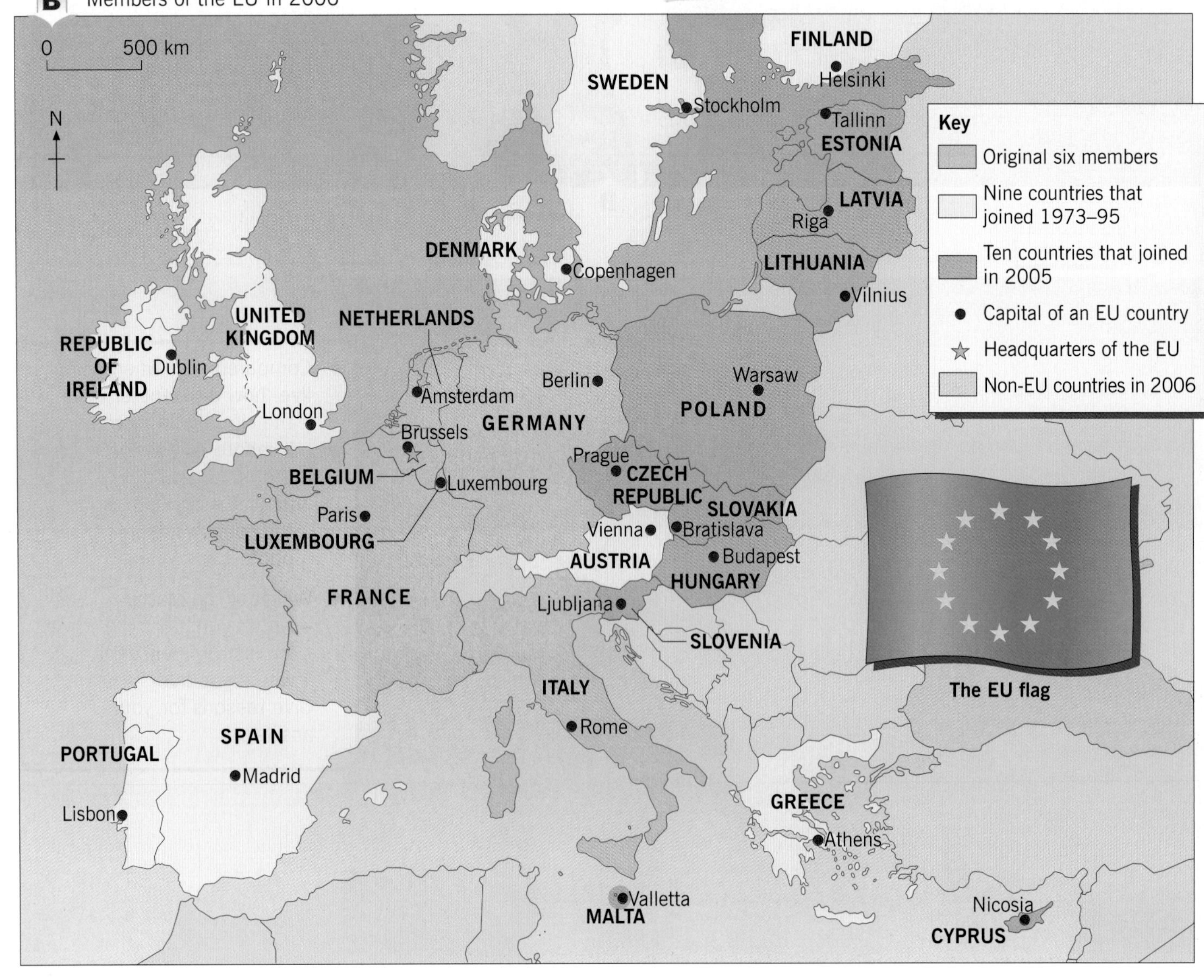

C What does the EU do?

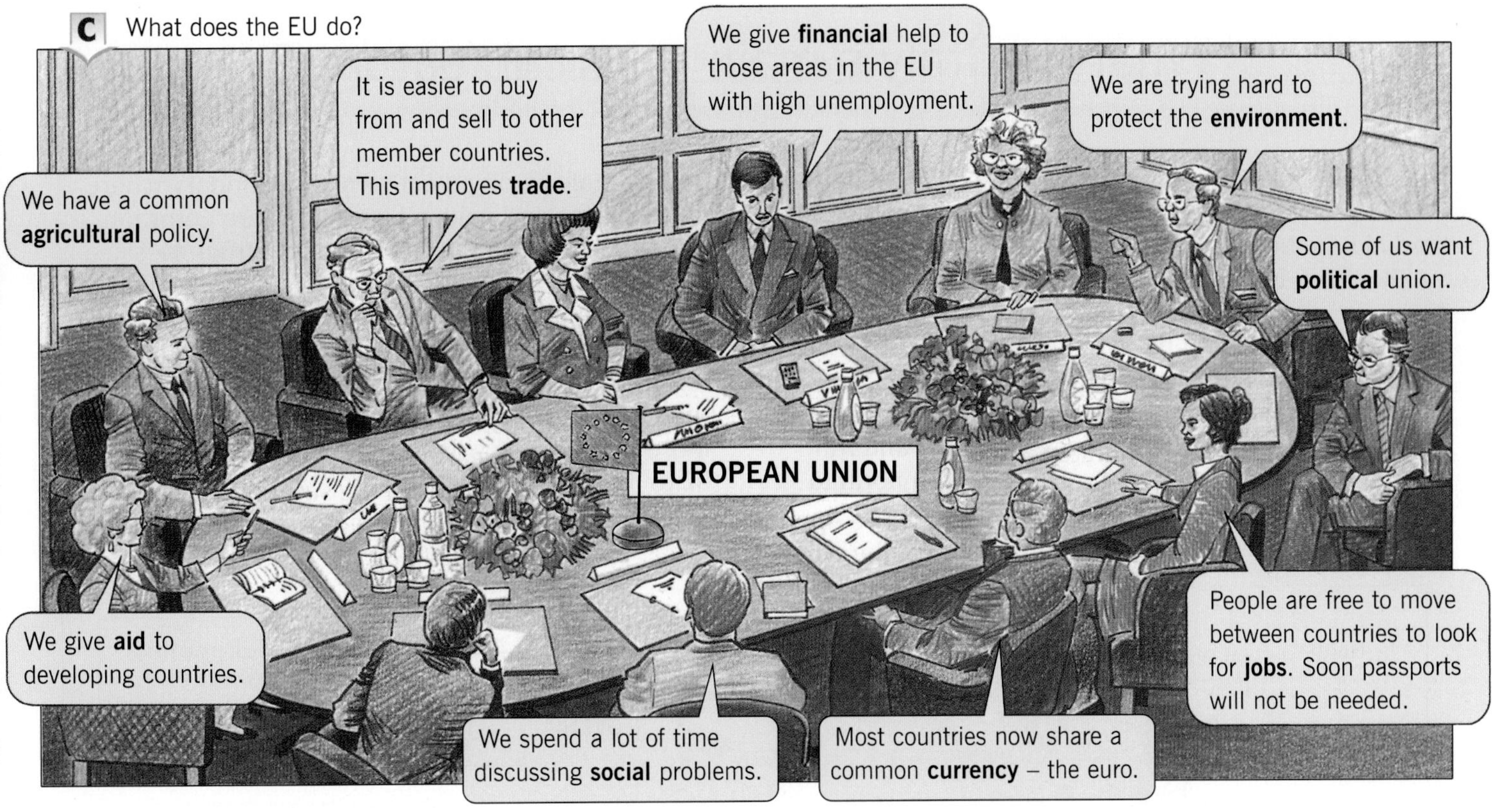

We have a common agricultural policy.
It is easier to buy from and sell to other member countries. This improves trade.
We give financial help to those areas in the EU with high unemployment.
We are trying hard to protect the environment.
Some of us want political union.
EUROPEAN UNION
We give aid to developing countries.
We spend a lot of time discussing social problems.
Most countries now share a common currency – the euro.
People are free to move between countries to look for jobs. Soon passports will not be needed.

Activities

1 **a** What is the European Union (EU)?

b When did the UK join the European Union?

c List six things that the EU does, according to diagram **C**.

2 **a** Make a copy of table **D**. In the second column, name:

- the first six member countries
- the nine that joined between 1973 and 1995
- the ten that joined in 2005.

b In the third column name the capital city for each of the 25 members.

D

	Name of country	Capital city
First 6 members		
Joined 1973–95		
Joined 2005		

3 Diagram **E** shows two groups of people with different opinions on the EU. Complete the speech bubbles to show:

a what one group sees to be the advantages

b what the other group sees as concerns.

E

We are in favour of the EU because ...
We have concerns about the EU because ...

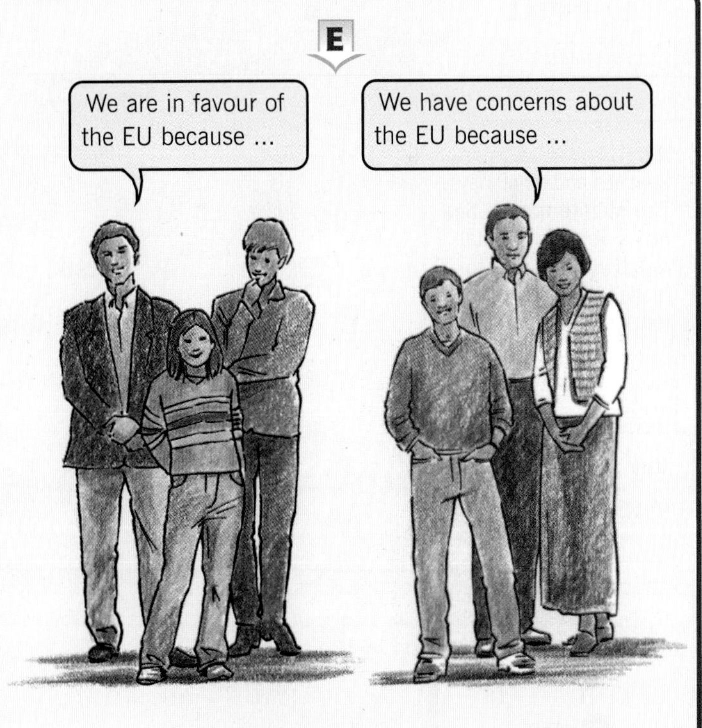

Summary

The European Union is, at present, a group of twenty-five countries working together to promote trade and economic, social and environmental policies.

What are Italy's main physical features?

Italy is a long, narrow country that stretches out into the Mediterranean Sea. The islands of Sicily and Sardinia are both part of Italy, as well as several smaller islands such as Elba and Capri.

Over three-quarters of the country is mountainous or hilly. The highest mountains are the Alps in the north. These have a permanent snow cover and glaciers fill the higher valleys. The Apennines are lower and run almost the whole length of the country from north to south.

Parts of Italy lie on an active plate boundary where earthquakes and volcanic eruptions are common. The world's fifth largest earthquake struck Messina in Sicily in 1908 killing 160,000 people. Italy has Europe's only active volcanoes. Mount Etna on Sicily and Vesuvius near Naples are the two most famous. Etna has erupted many times in recent years, whilst steam may often be seen rising from the summit of Vesuvius. More about these volcanoes may be found on pages 32 to 35.

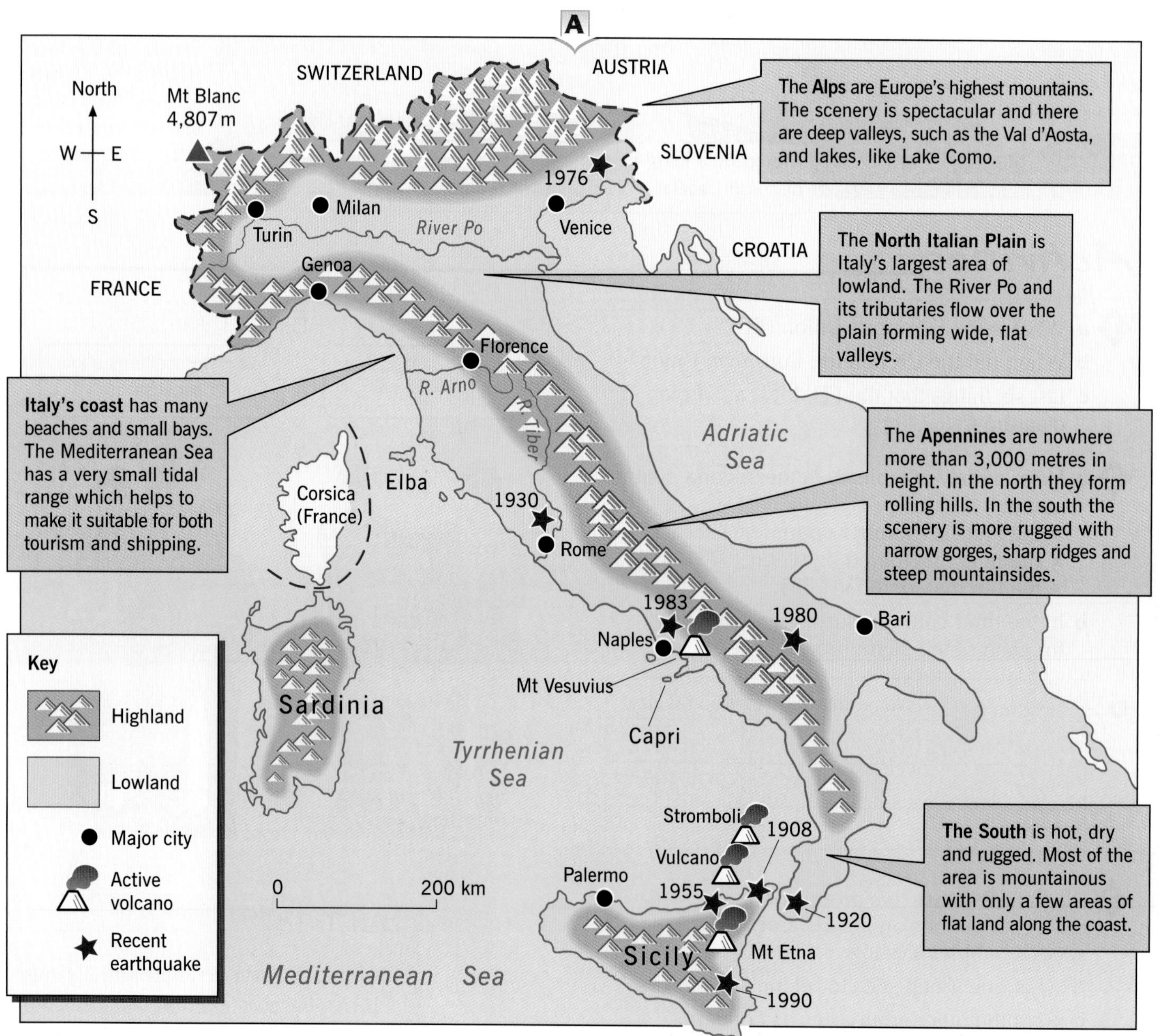

Most of Italy enjoys a pleasant **Mediterranean climate**, described on pages 16 and 17. The summers are hot and dry while the winters tend to be warm and wet.

Not all of the country is exactly the same, though. There are two main reasons for this. The first is that Italy is a very long country and stretches over 1,000 km from north to south. This means that the south is both warmer and drier than the north. In Sicily, for example, the summer drought may last for as long as four months and winter temperatures are, on average, a good 10°C higher than in the north.

The second reason is that Italy is a mountainous country. Temperatures usually decrease with height, so places in the Alps and Apennines are much colder than elsewhere. Villages in the mountain areas may be snow-bound for two or three months in winter.

B

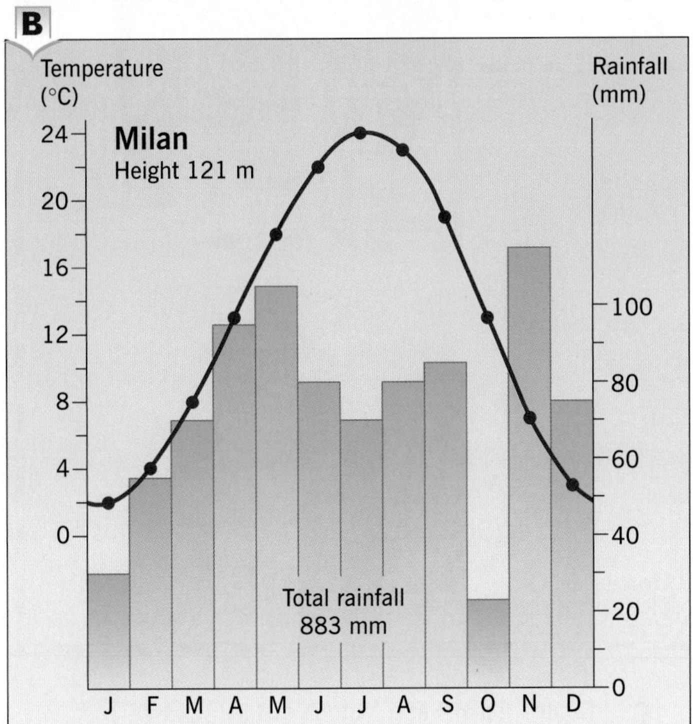

C

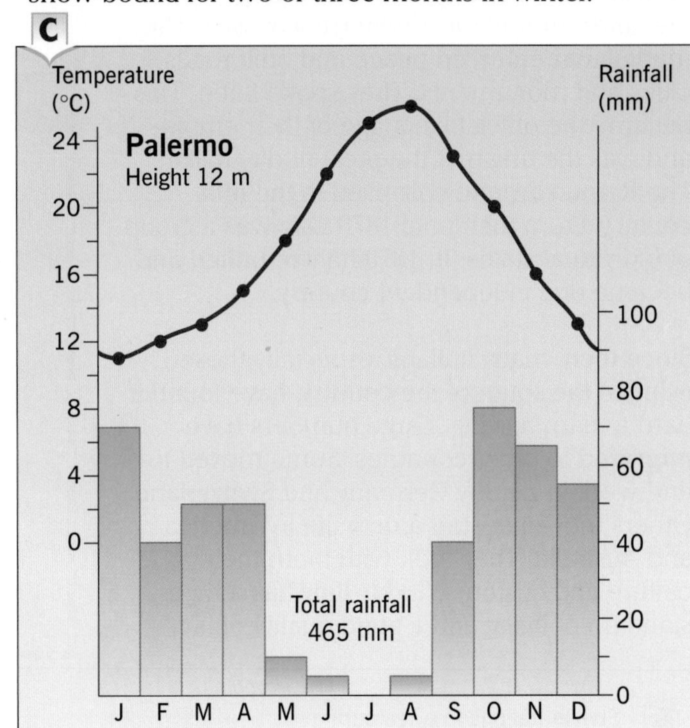

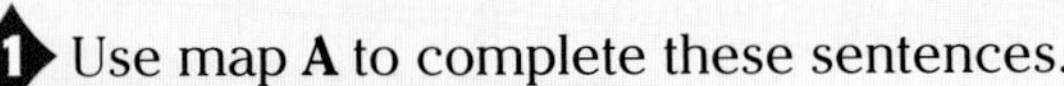

Activities

1 Use map **A** to complete these sentences.

a The highest mountains are ....

b The longest mountain range is ...

c Rome is on the River ...

d The longest river is the Po. It is ... km in length.

e The names of four volcanoes are ...

f The dates of the six most recent earthquakes are ...

g The hottest, driest and most rugged area is ...

h Three features of Italy's coast are ...

2 **a** Copy and complete table **D** to show the differences in climate between Milan and Palermo.

b Describe the climates of Milan and Palermo by using the correct statements from drawing **E**.

3 Explain why Palermo is warmer than Milan. The information on page 6 will help you.

D

		Milan	Palermo
Temperature	January		
	July		
Rainfall	January		
	July		

Summary

Italy is a mountainous country with little flat land. The climate varies from place to place but most areas have long, hot summers, warm winters and plenty of sunshine. Winters can be wet.

Who are the Italians?

Two thousand years ago the country that is now called Italy was part of the Roman Empire, shown in map **A**. The Romans brought civilisation to much of Western Europe and the lands around the Mediterranean Sea. They made laws, enforced peace and built roads, cities and monuments. They spoke Latin. This became the official language of their empire and was the origin of the present-day Italian. The Roman Empire collapsed in the fifth century. From then until 1870 Italy was a group of individual states. In 1870 Italy re-united and became one independent country.

Since then, many Italians, especially those living in the south of the country, have found it hard to earn a living. Large numbers have **migrated** to other countries. Some moved to find work in nearby Germany and Switzerland. Others moved to start a new life in America and Australia. They took with them their culture and customs. Today Italy has a high standard of living and a good quality of life.

A

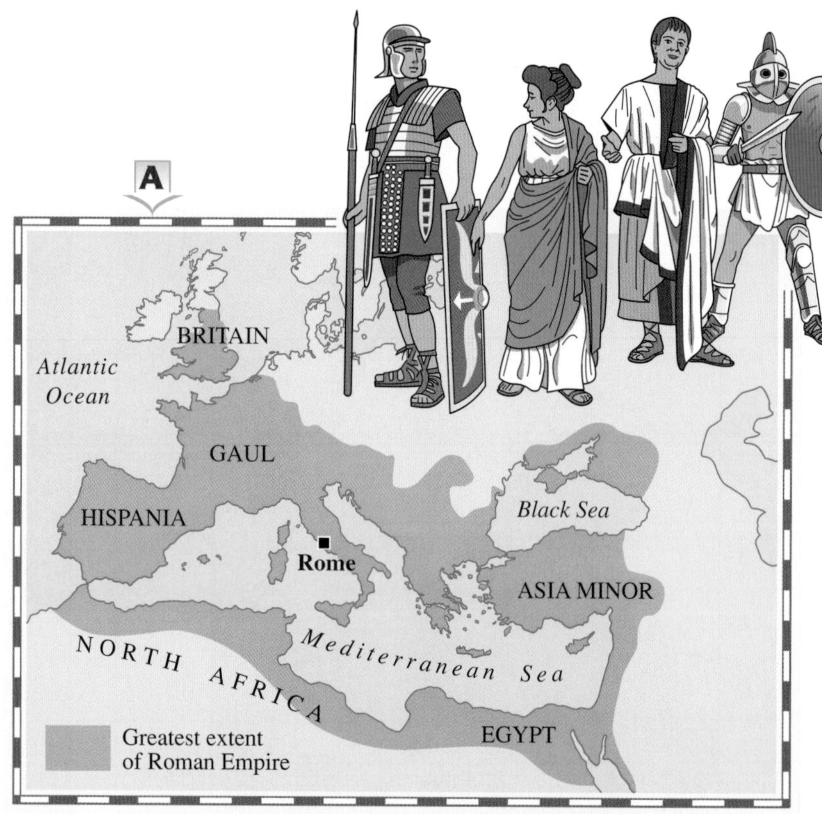

B Stylish Italians in a restaurant

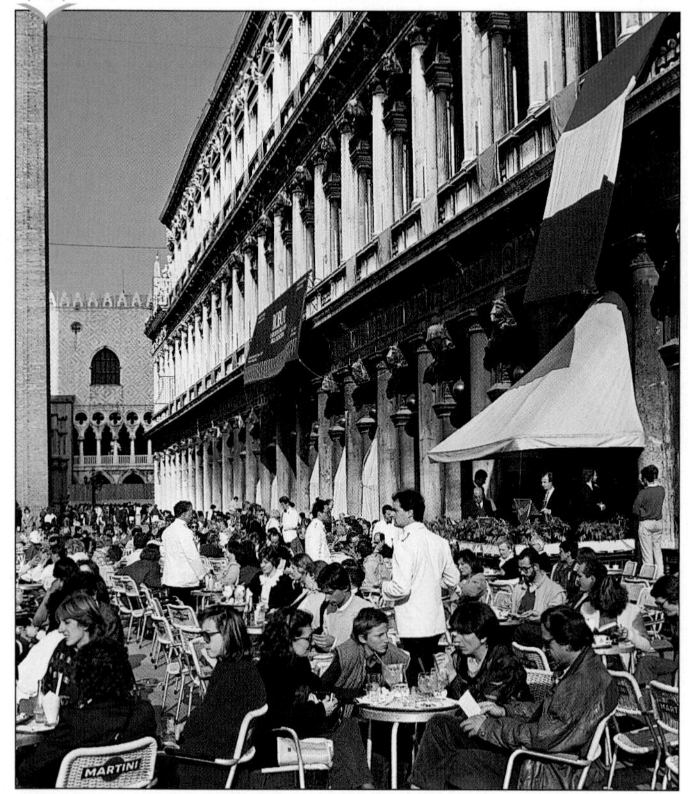

C Excited Italians at a football match

Different groups of people tend to develop their own customs and way of life. This can include their language, music and religion, how they dress and behave, what they eat and drink and what they do in their spare time. This creates a **sense of identity**.

The photos in **D** show some ways in which Italians have developed a sense of identity. We must remember, however, that there are many types of Italians. Not every Italian will be the same as those shown below.

D

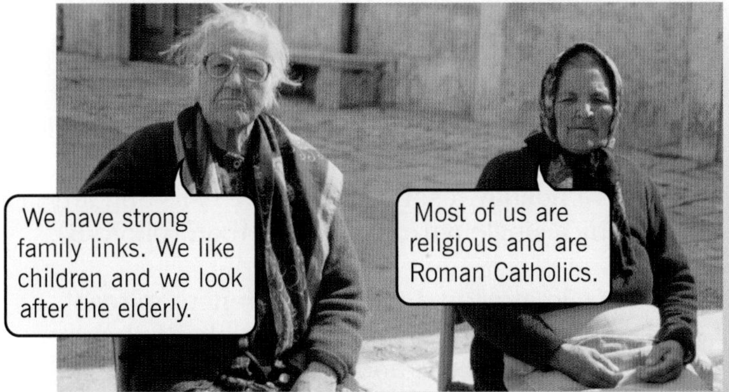

1 **a** What did the Romans bring to much of Europe?

b In what ways have modern Italians created their own customs and way of life?

2 **a** Why have many Italians migrated since independence in 1870?

b To which countries did they migrate?

3 **a** Using the photos in **D**, suggest how members of an Italian family might spend a weekend.

b Using diagram **E**, make up a menu for a typical Italian meal.

MENU

• Starter	• Drink during meal ....
• Main dish	• Drink after meal ....
• Dessert	

E

WINES: Chianti, Lambrusco, Marsala, Valpolicella
Olive oil
CHEESES: Gorgonzola, Parmesan
PASTAS: Spaghetti, Tortelloni, Ravioli, Lasagne, Macaroni
Minestrone, Pizza, Salami
FRUIT: Oranges, Lemons, Tomatoes, Grapes, Olives
Ice cream
COFFEE: Cappucino, Espresso

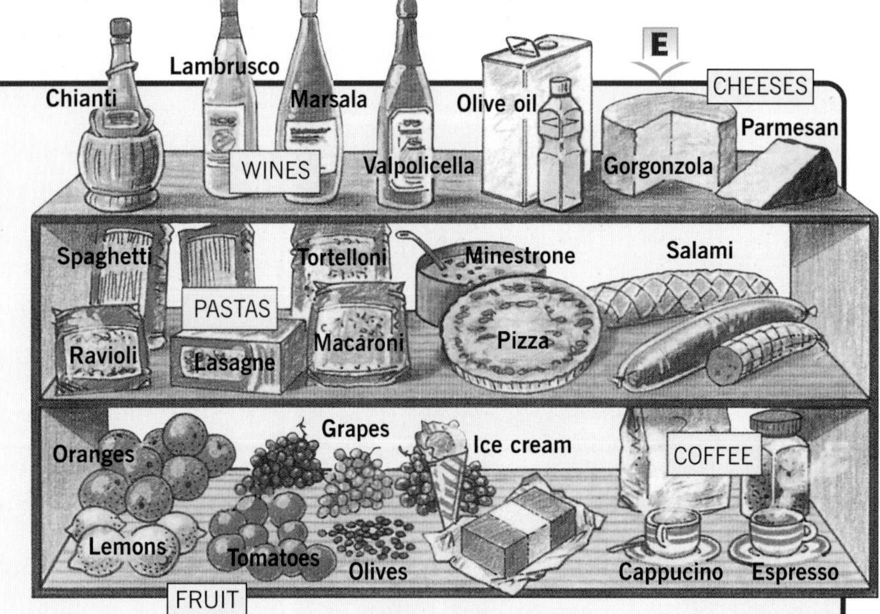

Summary

Italy is a country with a long and interesting history. Over the years Italians have developed their own customs and way of life.

How does the environment affect people?

One of the earliest settled parts of Italy was around Mount Vesuvius and the Gulf of Naples (map **A**, page 90 and photo **A** on page 86). The soil around the mountain was perfect for farming and the sea allowed people to trade. It was not until Vesuvius erupted in AD79 and destroyed the towns of Pompeii and Herculaneum (sketch **A**) that people realised it was a volcano. The ruins of these Roman towns and the Mediterranean climate now attract many tourists (page 16). Apart from tourism, most people find jobs as farmers or work in the large city of Naples.

To the south, the Sorrento **peninsula** is very different. The rock is limestone which gives hilly land and poor soil. Most of the area is covered in typical Mediterranean scrub vegetation (page 18). Along the south coast there is hardly any flat land. Hillsides have been terraced for farming. Several small, attractive holiday resorts cling to the cliffs and are linked by a single narrow, twisting road (photo **B**).

It is the climate and the local differences in relief and soils which are mainly responsible for the occupations, land use and settlement pattern of the area.

C
An **infrared** photo taken from a **Landsat** satellite as it circles the earth at a height of 900 km

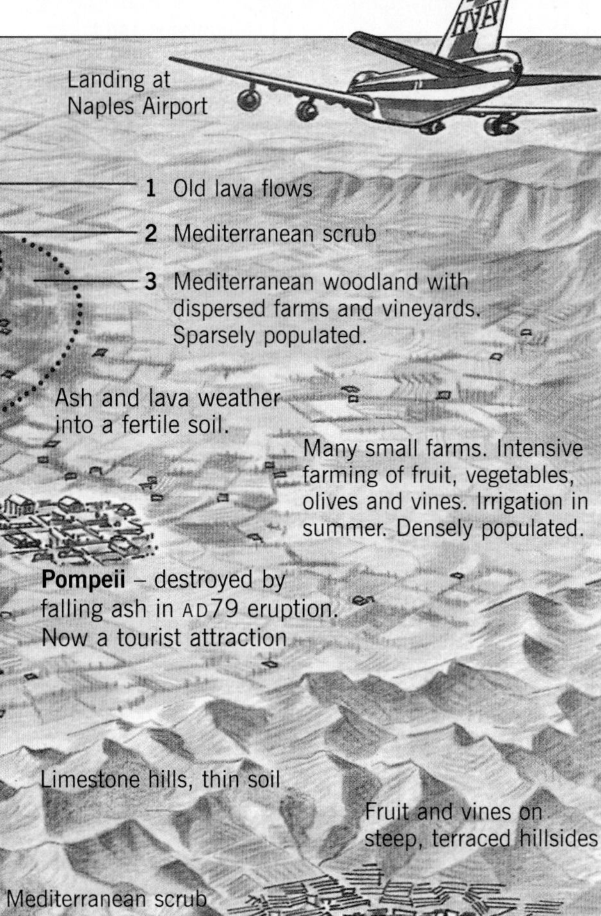

Activities

1 The area shown on sketch **A** can be divided into four parts. These have been labelled **W**, **X**, **Y** and **Z** on map **D**.

a Make a larger copy of table **E** and complete it by using the information given on these two pages.

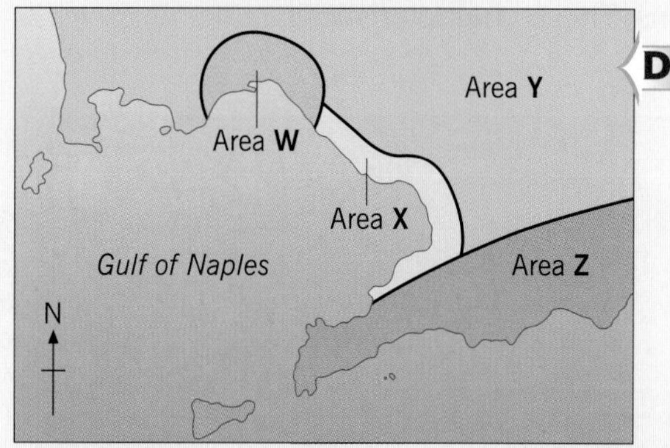

E

	Area W	Area X	Area Y	Area Z
Settlement e.g. dispersed, linear, nucleated				
Land use e.g. built-up, farming, scrub				
Jobs e.g. tourism, farming, factories				

b Give two reasons why farming is better in Area **Y** than Area **Z**.

c Suggest why the settlement patterns in Areas **X** and **Z** are different.

2 A Landsat photo shows false colours. Using photo **C** and sketch **A**, match the following Landsat colours with the correct type of land use.

Summary

The main occupations, land use and settlement patterns of a locality can often be explained by its environment.

What are Italy's main regions?

What is a region? The term **region** has been used several times without explaining what it means. A region is an area of land which has common characteristics. It is therefore different from other regions. These characteristics can be:

- **Physical**, where places have the same climate, vegetation or soils.
- **Human** and **economic**, such as political areas or places with the same economic activities or level of development.

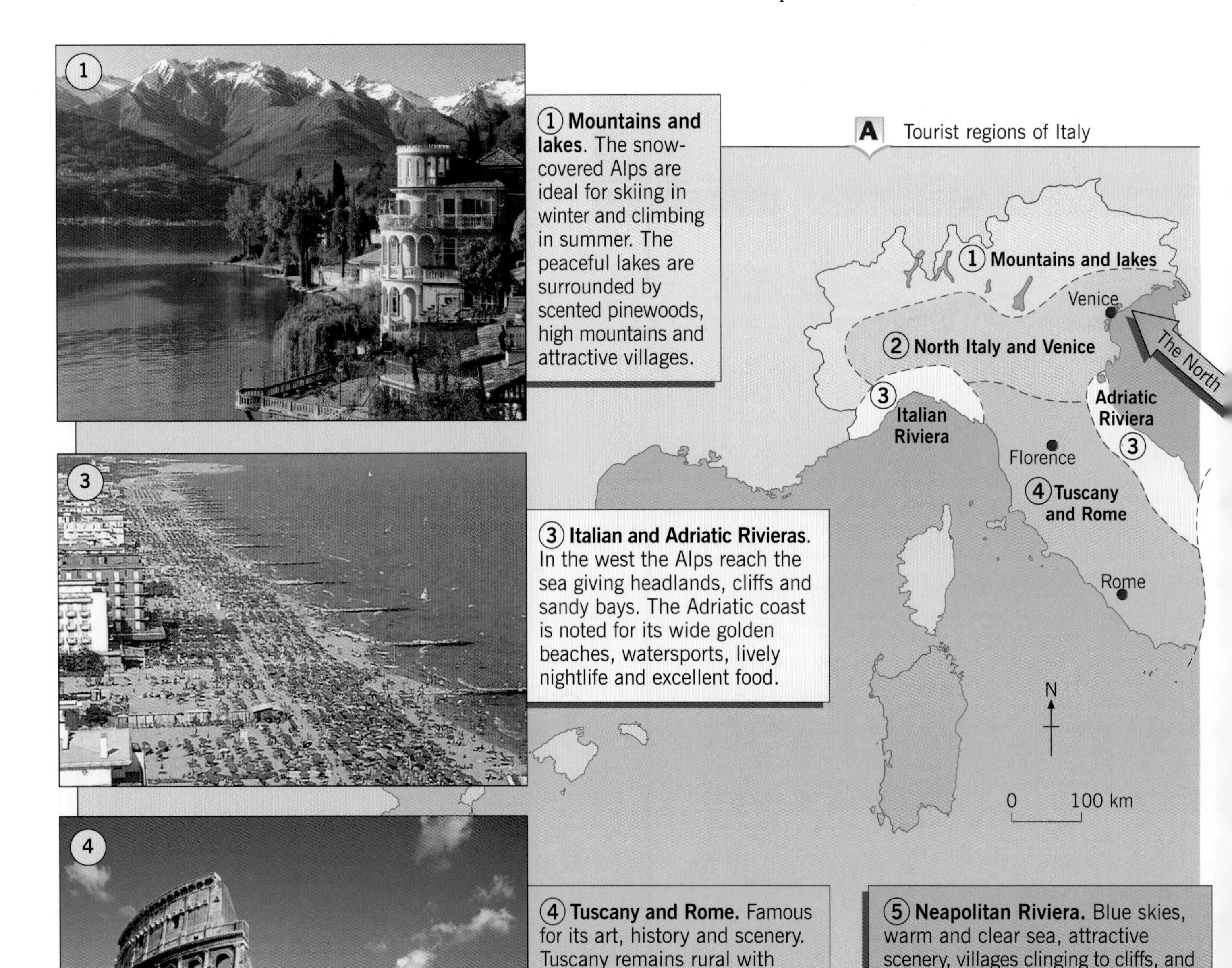

1 **Mountains and lakes.** The snow-covered Alps are ideal for skiing in winter and climbing in summer. The peaceful lakes are surrounded by scented pinewoods, high mountains and attractive villages.

3 **Italian and Adriatic Rivieras.** In the west the Alps reach the sea giving headlands, cliffs and sandy bays. The Adriatic coast is noted for its wide golden beaches, watersports, lively nightlife and excellent food.

A Tourist regions of Italy

1 Mountains and lakes
2 North Italy and Venice
3 Italian Riviera
3 Adriatic Riviera
4 Tuscany and Rome
Venice
Florence
Rome
The North
N
0 100 km

4 **Tuscany and Rome.** Famous for its art, history and scenery. Tuscany remains rural with wooded hills and large vineyards. Rome and Florence have numerous art galleries, historic buildings, fine shops and excellent food.

5 **Neapolitan Riviera.** Blue skies, warm and clear sea, attractive scenery, villages clinging to cliffs, and historic ruins (pages 78–79).

6 **The South.** Inland there are mountains and quiet, unspoilt villages. The coastal white, sandy beaches are quiet. A place to relax (pages 86–87).

What are Italy's regions? Different people using different characteristics can produce different maps showing the regions of Italy. There is no single correct map. For example:

- There are three main **physical** regions (map **A** page 90). These are the Alps, the North Italian Plain and the Apennines (peninsular Italy).
- There are twenty states. These **political** regions are similar to counties in Britain.
- There are two **economic** regions – the more developed North and the less developed South. The remainder of this unit describes how these two economic regions are different.
- There are seven tourist regions which are shown on these two pages.

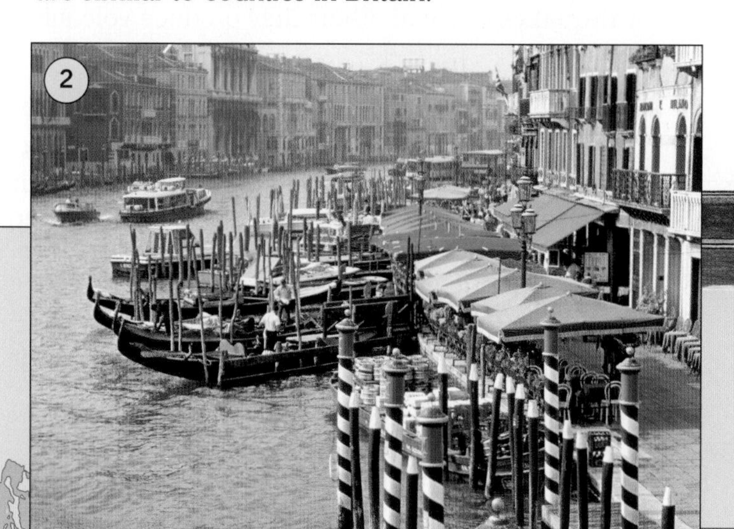

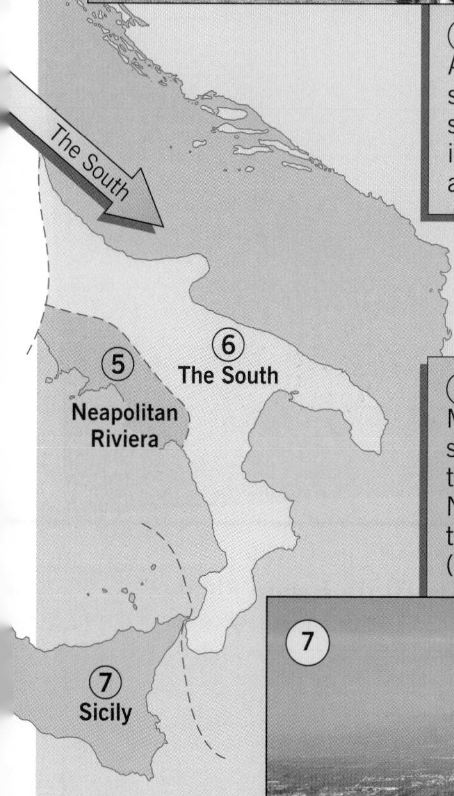

② North Italy and Venice. A lowland area which has many small historic towns with their old squares and churches. In the east is Venice with its numerous canals and, nearby, long sandy beaches.

⑦ Sicily. The largest island in the Mediterranean. Days are usually sunny. There are remains of Greek temples, Roman theatres and Norman castles. The highlight of the attractive scenery is Mount Etna (pages 38–39).

Activities

1 **a** What is a region?
b What is the difference between a physical and a political region?
c Name three physical regions in Italy.
d Name the two economic regions in Italy.

2 Which region, or regions, would you visit in Italy if you wanted to:
a visit art museums
b ski
c travel by boat on canals
d see a volcano
e lie in the sun on a sandy beach
f climb mountains
g enjoy nightlife
h take part in watersports
i see quiet villages
j visit historic buildings
k visit a vineyard?

3 Which of the seven tourist regions shown on map **A** would you most like to visit? Give reasons for your answer.

Summary

A region is an area of land which has common characteristics. These characteristics, which can be physical or human and economic, make each region different from other regions.

Physical features of North and South Italy

Landforms

The **North Italian Plain** was once part of the Adriatic Sea and lies between the Alps and the Apennines. Fast-flowing rivers from those highland areas brought down **silt** which they deposited in the shallow sea. The silt built up above sea-level forming a very flat and fertile **plain**. When snow in the Alps melts in spring the rivers may flood this low-lying land. The main river, the Po, is continually pushing its **delta** outwards into the Adriatic. Embankments have been built to try to stop it overflowing onto its flood plain. Little natural vegetation remains on the plain. Lombardy poplar trees have been planted to try to reduce the force of the wind.

A The North Italian Plain

The **South** was once part of the Mediterranean Sea. Rocks were formed on the sea bed and later pushed up to form the Apennines. These earth movements still occur today. Sometimes they cause serious **earthquakes** while at others they produce volcanic eruptions.

The steep-sided Apennines were once covered in Mediterranean woodland. When this woodland was cleared the soil was exposed to the heavy winter rain and washed away. Many parts are now either eroded (photo **B**) or covered in scrub vegetation (page 18). There is not much fertile land. The best soils are:

1 Where rivers have deposited silt as deltas at their mouths. (Most rivers are seasonal and only flow in winter.)

2 Near to volcanoes where the lava and ash soon weather into a deep soil.

B Soil erosion in the Apennines

Activities

1 Landsketches **F** and **G** show parts of the North Italian Plain and the Apennines. On large copies of these sketches put the following labels in the correct places.

poplar trees | scrub vegetation | fertile silt | flat plain | steep hillsides | thin soil | permanent river | seasonal river | possible flooding | possible earthquake

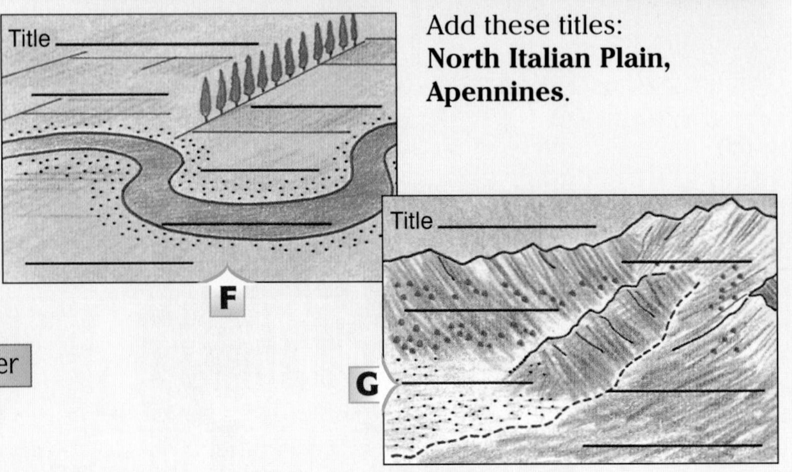

Add these titles: **North Italian Plain, Apennines.**

Climate

As we have seen on pages 92 and 93, most of Italy has a **Mediterranean climate** with hot dry summers and warm wet winters. Although the climates of the North Italian Plain and the South of Italy are largely the same, there are also some differences. These are given in table **C**.

	North Italian Plain	The South
Type of climate	Between a British and a Mediterranean climate.	Mediterranean climate (page 16)
Winter temperatures	Cold. January is between 0° and 2°C.	Warm. January is between 8° and 10°C.
Summer temperatures	Very warm. 24°C in July.	Hot. Over 26°C in July.
Rainfall	800 mm spread evenly throughout the year. Rain is usually not very heavy.	700 mm which nearly all falls in winter. Rain is usually heavy. Very little rain in summer.
Hazards	Frost is common in winter. Fog is very common at any time of year but mainly in winter (photo **D**). Milan averages 100 days of fog a year.	Drought and heatwave conditions in summer (photo **E**). Snow on higher slopes in winter.

C

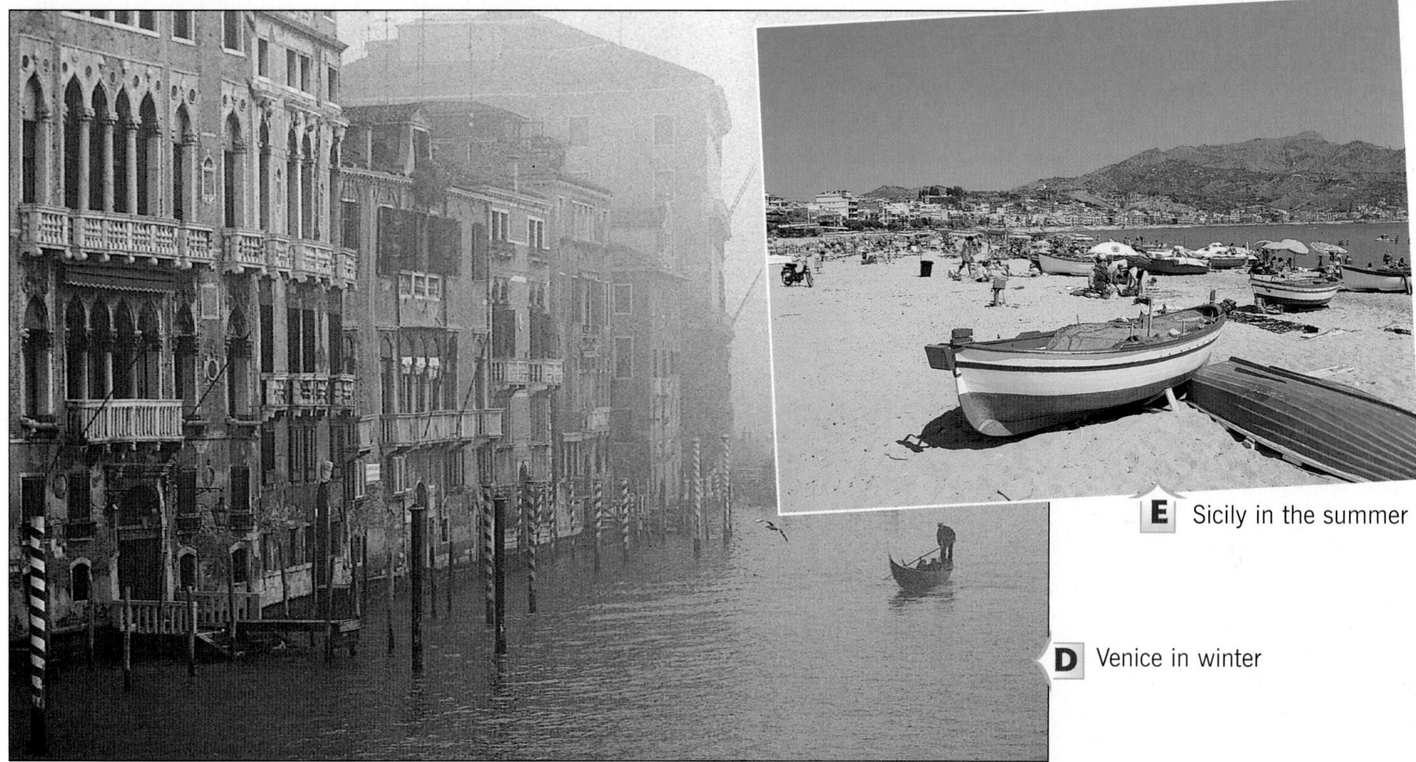

E Sicily in the summer

D Venice in winter

2 Make a copy of map **H**. Add the labels by choosing the correct word from the pair in brackets to show the changes as you move north or south.

3 Give three differences between the climate of the North Italian Plain and the South of Italy in:
- summer
- winter.

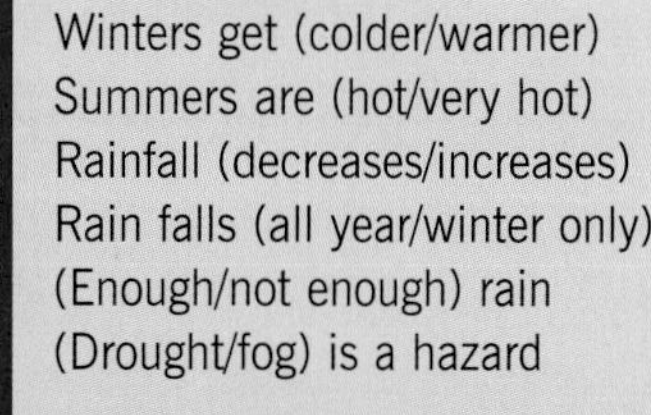

H

Winters get (colder/warmer)
Summers are (hot/very hot)
Rainfall (decreases/increases)
Rain falls (all year/winter only)
(Enough/not enough) rain
(Drought/fog) is a hazard

Summary

There are more differences than similarities between the landscapes and climate of the North Italian Plain and the South of Italy.

Life on the North Italian Plain

The North Italian Plain is the richest region in Italy. It has a standard of living as high as anywhere in the EU. Most people live in large industrial towns and cities. Industry has attracted large numbers of workers since 1950. Some of these workers came from rural areas of the North Italian Plain (rural–urban migration) but most arrived either from the South of Italy or from poorer countries which surround the Mediterranean Sea.

Land use and jobs

Although only 4 in every 100 workers are farmers, agriculture is still a major type of land use (graph **A**). Most farms are large and the fields are grouped together, making them easier for the farmer to reach. Farming is **intensive**, meaning that no land is wasted, and **commercial**, which means that farm produce is sold for a profit. Vines are grown in the west of the region; fruit, wheat and rice in the centre; and maize (corn) in the east.

A Jobs on the North Italian Plain

Primary **4%**
Tertiary (services) **52%**
Secondary **44%**

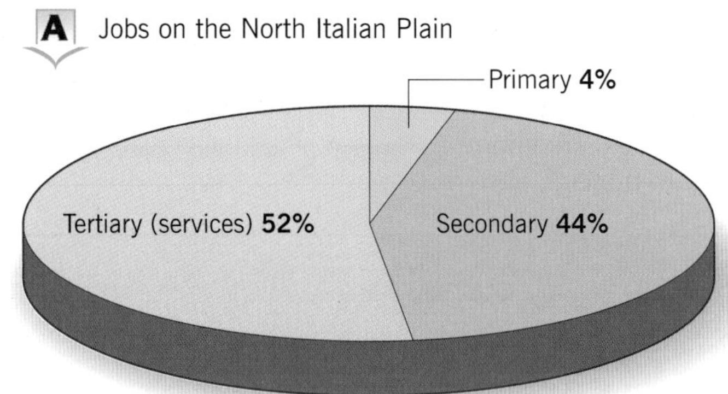

B Advantages of the North Italian Plain for farming

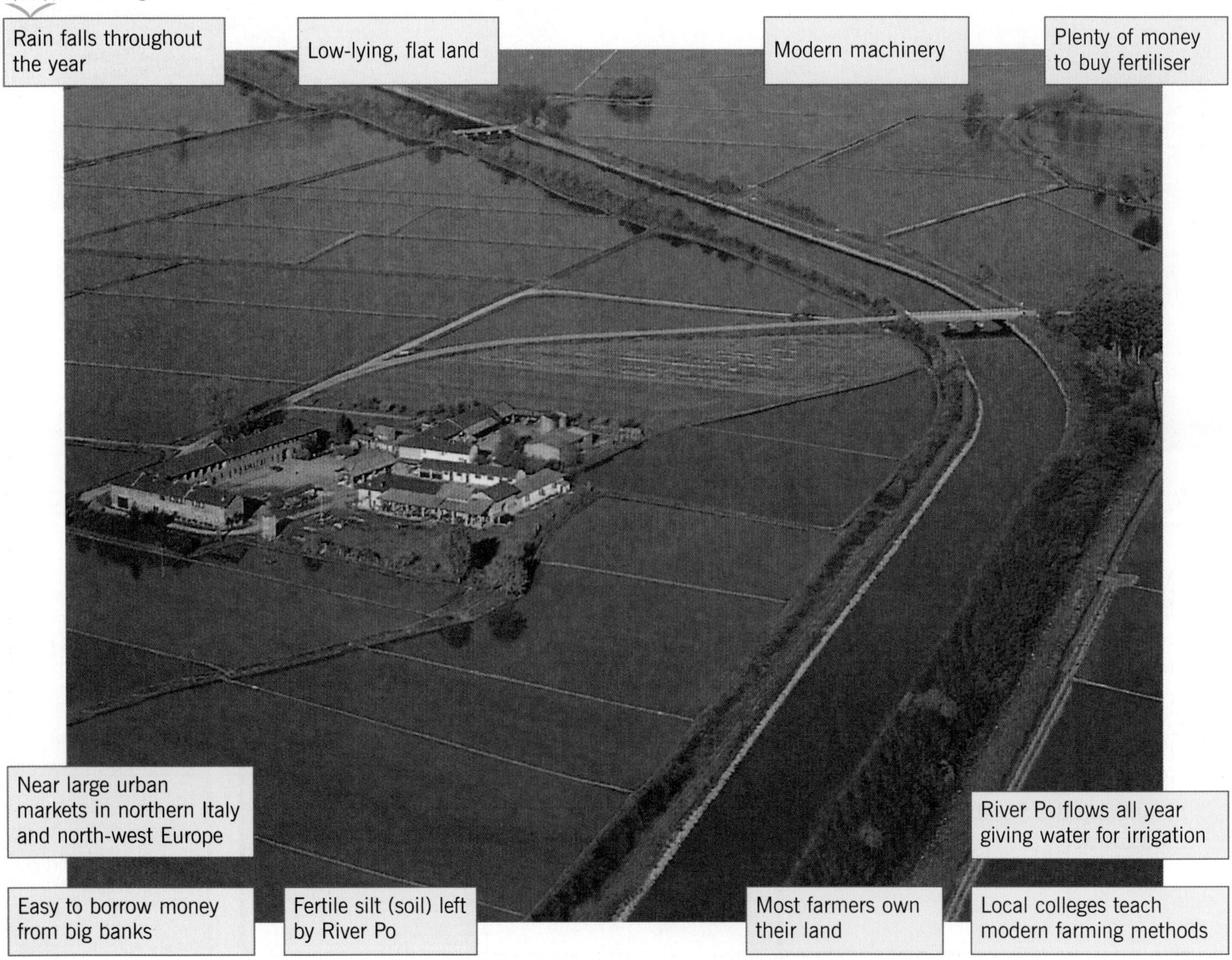

Rain falls throughout the year
Low-lying, flat land
Modern machinery
Plenty of money to buy fertiliser
Near large urban markets in northern Italy and north-west Europe
River Po flows all year giving water for irrigation
Easy to borrow money from big banks
Fertile silt (soil) left by River Po
Most farmers own their land
Local colleges teach modern farming methods

Industry

The west and centre of the North Italian Plain has always been the most important industrial region in Italy. Although at first centred on the '**industrial triangle**' between the cities of Turin, Milan and Genoa, it is now spreading outwards into the surrounding regions (map **D**).

The most important firm is Fiat whose large car assembly plant is in Turin (photo **C**). The present head of Fiat also owns Turin's daily newspaper and Juventus Football Club. Milan is Italy's largest city and the centre for banking and fashion. Much of the region's trade either has to pass through the port of Genoa or has to cross the Alps.

Although this is the richest part of Italy with most of the better paid and skilled jobs, it still has its problems. As more and more land is being built upon there is less for farming and recreation. Towns have grown so quickly that there has not been enough time to plan them carefully. Houses and flats have been built very close together, roads are congested and there is very little open space and parkland. There has been little care of the environment and many pollution problems have developed.

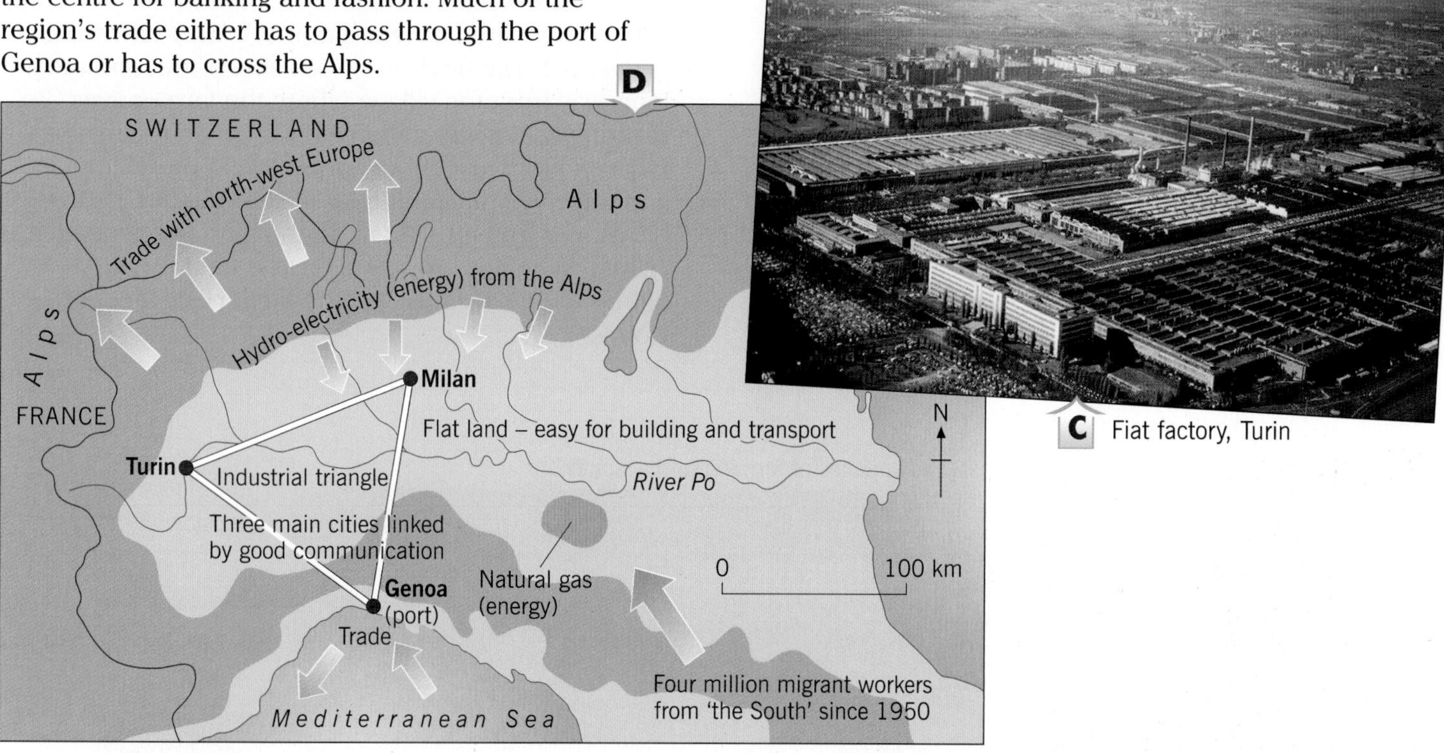

C Fiat factory, Turin

Activities

1. Make a large copy of graph **E**. Add the information from graph **A** to show the main types of jobs on the North Italian Plain.

2. **a** List six advantages of farming in this region.
   b Sort the advantages under the headings **Physical** and **Human**.

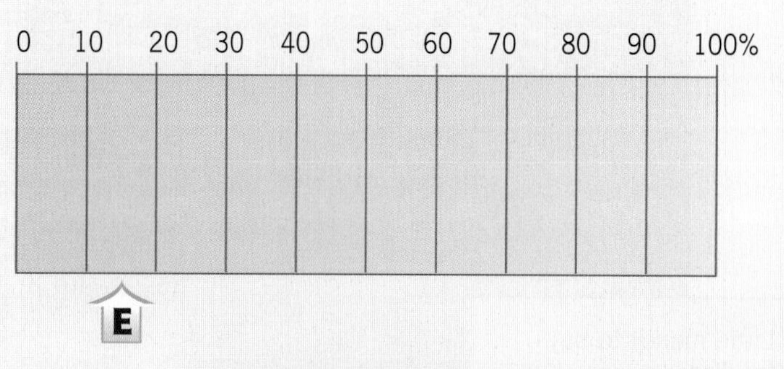

3. **a** Name four important industries in this region.
   b Give three reasons why this is the most important region for industry in Italy.
   c Draw a star diagram to show some of the problems which have resulted from industrial growth.

Summary

The North Italian Plain is the richest region in Italy. Most of the land is either used for large towns and industry or for commercial farming.

Life in the South of Italy

The South of Italy is the poorest region in Italy. In 2005, one part of it had the lowest standard of living in the EU. Most people still live in hilltop villages in rural areas (photo **B**). The few towns, which are on the coast, have little industry to attract people.

Many people from the South have had to migrate either to the North of Italy, to other EU countries or even to North America or Australia to find work. Recently even poorer people from south-east Europe and developing countries have moved into this region.

A Jobs in the South of Italy

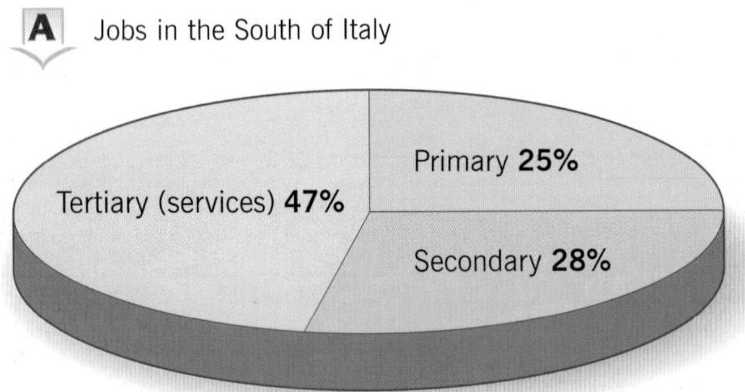

Land use and jobs

Agriculture is still the major type of land use and 25 in 100 workers are farmers (graph **A**). Most farms are very small. The fields are often spread out and are a long way from the village where the farmer lives. Farming is at a **subsistence** level which means that farmers grow just enough food for their own needs and have very little left over to sell (photo **C**). Vines, olives and fruit are grown on the hillsides beneath the village. Wheat is grown where the land is flatter. Sheep and goats graze on the higher and steeper slopes.

B Disadvantages of farming in the South

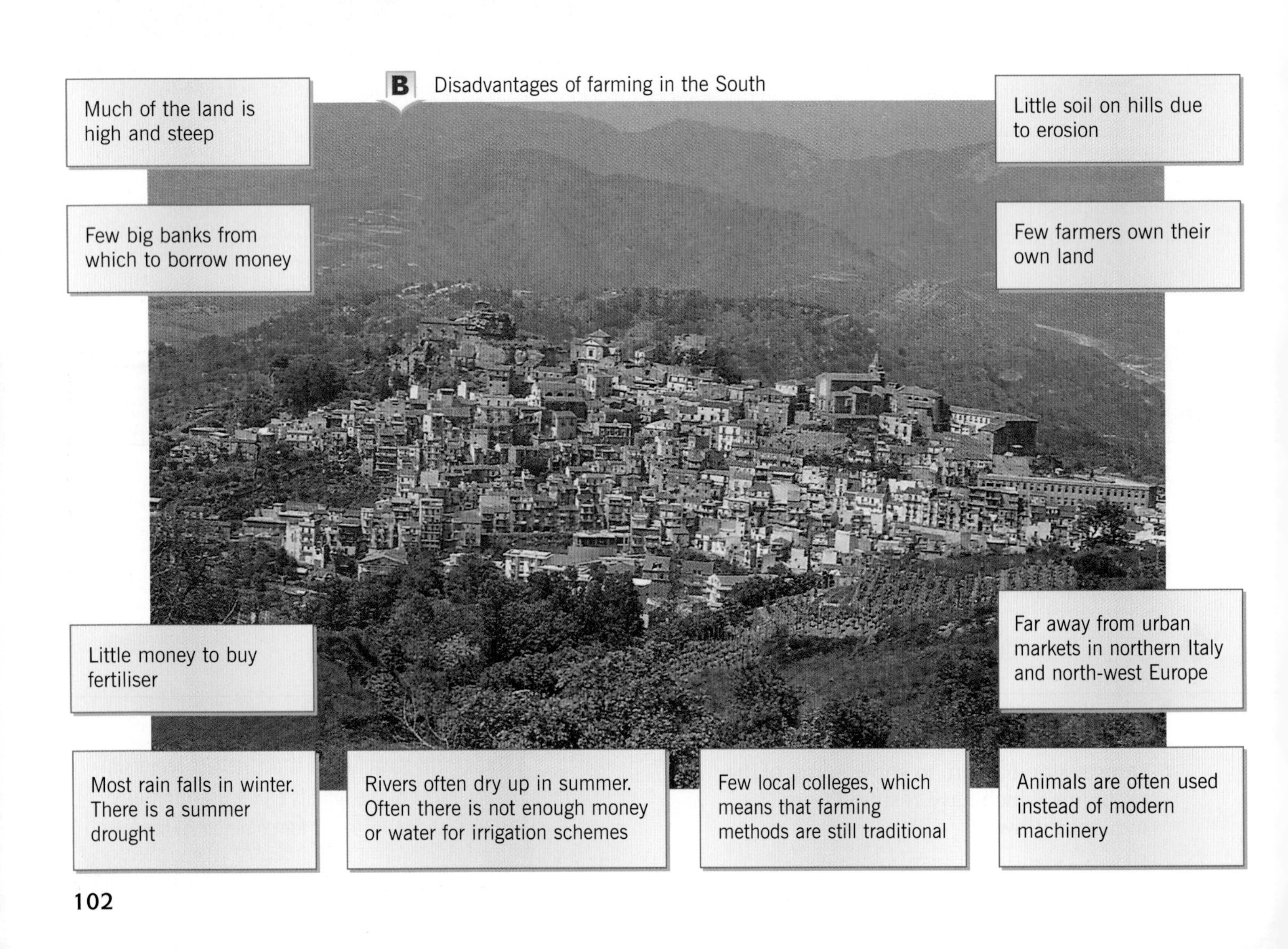

Much of the land is high and steep

Little soil on hills due to erosion

Few big banks from which to borrow money

Few farmers own their own land

Little money to buy fertiliser

Far away from urban markets in northern Italy and north-west Europe

Most rain falls in winter. There is a summer drought

Rivers often dry up in summer. Often there is not enough money or water for irrigation schemes

Few local colleges, which means that farming methods are still traditional

Animals are often used instead of modern machinery

Industry

The South of Italy has never been an important industrial region. It is isolated from the rest of Europe by poor transport links. The area has very few natural resources, little money and limited skills. The high birth rate means there are too many people looking for few available jobs. Large numbers, therefore, have had to leave the region in order to find work.

Attempts have been made to improve roads and to introduce new industries. This has only been partly successful as the few new industries, such as steelmaking, chemicals and car assembling, have only benefited a few places (map **D**). Also, as these industries are highly mechanised they do not need to employ many people.

Despite these problems the South is slowly becoming better off. Marshy areas have been drained and trees planted. New dams, irrigation schemes and motorways have been built. The hot, dry summers and sandy beaches are attracting more tourists. Some of the earlier migrants to the North have returned with the money which they earned there. Even so, the gap in wealth between the North and the South of Italy continues to grow.

C The South of Italy

D

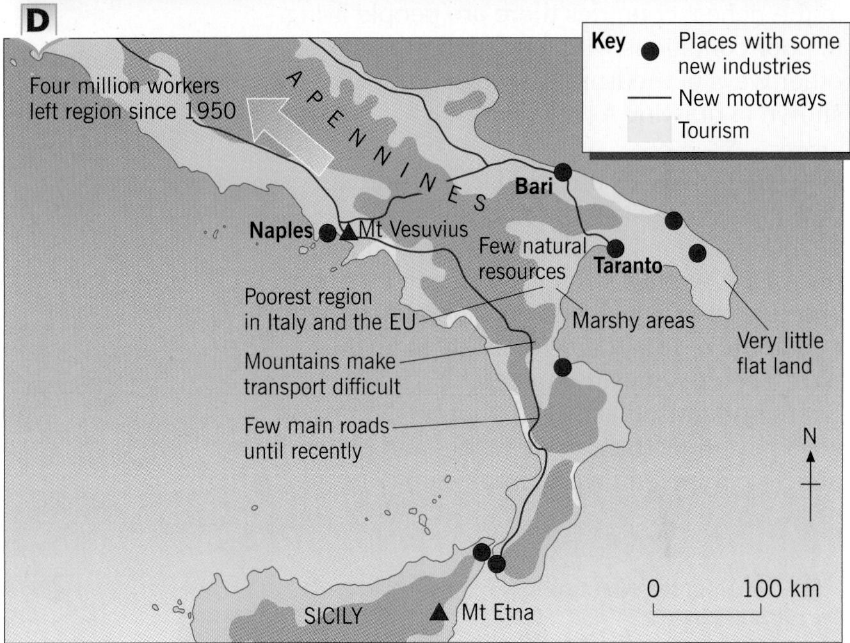

Activities

1. Make a large copy of graph **E**. Add the information from graph **A** to show the main types of jobs in the South of Italy.

2. **a** List six problems facing the farmers in this region.
   b Sort these problems under the headings **Physical** and **Human**.

3. **a** Name three important industries in this region.
   b Give three reasons why this area has had difficulty in attracting industry.
   c Draw a star diagram to show some of the recent improvements made in this region.

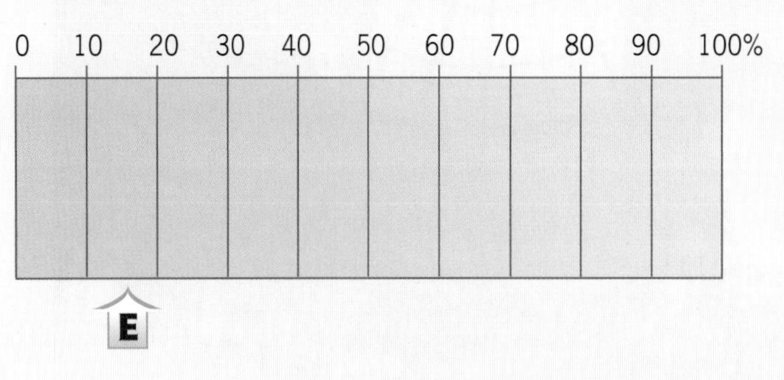

E

Summary

The South of Italy is the poorest region in the country. Most of the land is used for subsistence farming and some for industry and tourism. Large areas have very limited use for people.

How developed is Italy?

All countries are different. Some, like the UK, are wealthy and have a **high standard of living**. They are said to be **developed**. Others, like Kenya and India for example, are poor, have a low standard of living and are said to be **developing**.

Development is about improving the quality of life for people, but measuring the actual level of development can be difficult. The most commonly used method is to look at wealth. There are problems with this, though, as even in the richest countries there are people living in poor conditions with little money. Some other ways of measuring development are shown in drawing **A**.

Whatever methods are used to measure development, Italy is certainly a highly developed country. Indeed economically it is one of the world's seven richest nations and in recent years has enjoyed considerable growth and improved standards of living. Industries have been particularly successful. They have been modernised and with improved transport links have become more efficient and better able to compete in world markets.

A Some ways of measuring development

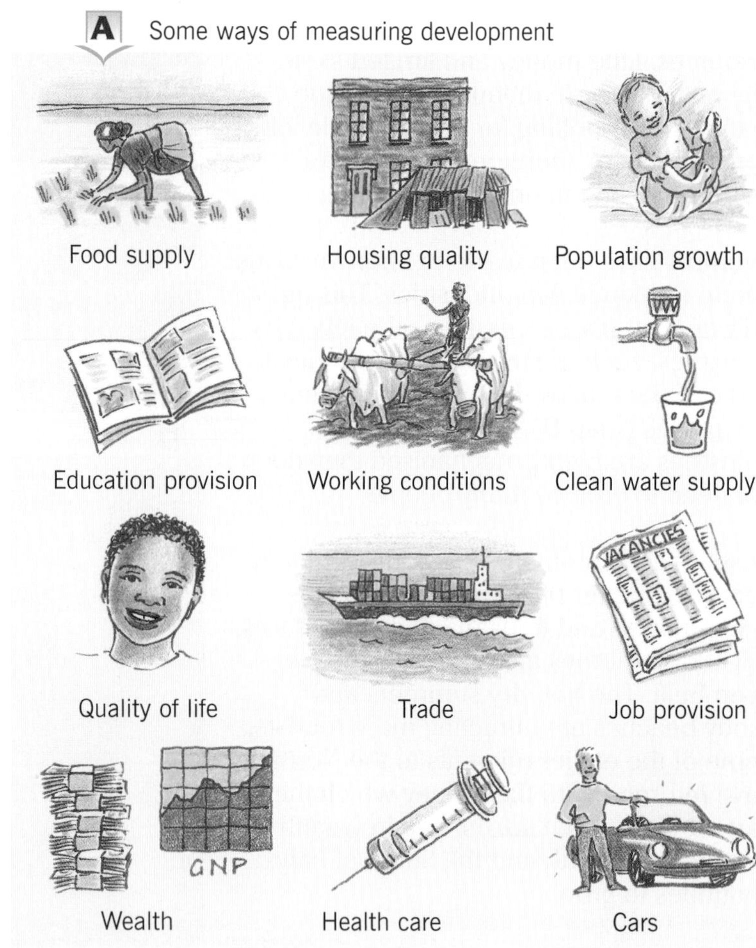

B Milan in the North of Italy

C A farm worker in the South of Italy

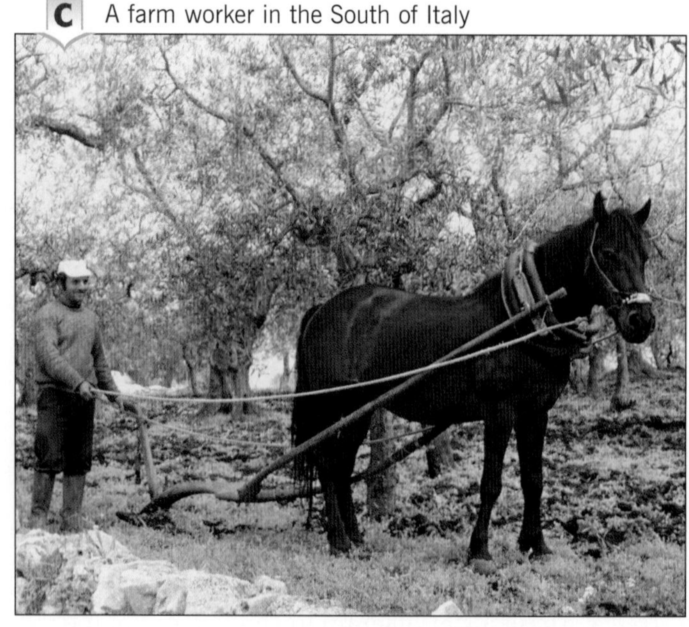

As photo **C** suggests, however, not all of Italy is wealthy and highly developed. As we have seen on pages 98 to 103, the physical features, way of life and economic progress of the North are very different from those of the South. The North is much wealthier than the South and people in the North enjoy high living standards and a good quality of life. The South is less well off and, despite considerable government action and several improvement plans, many people in that region are still poor, live a difficult life and have a low standard of living.

D The North and South compared

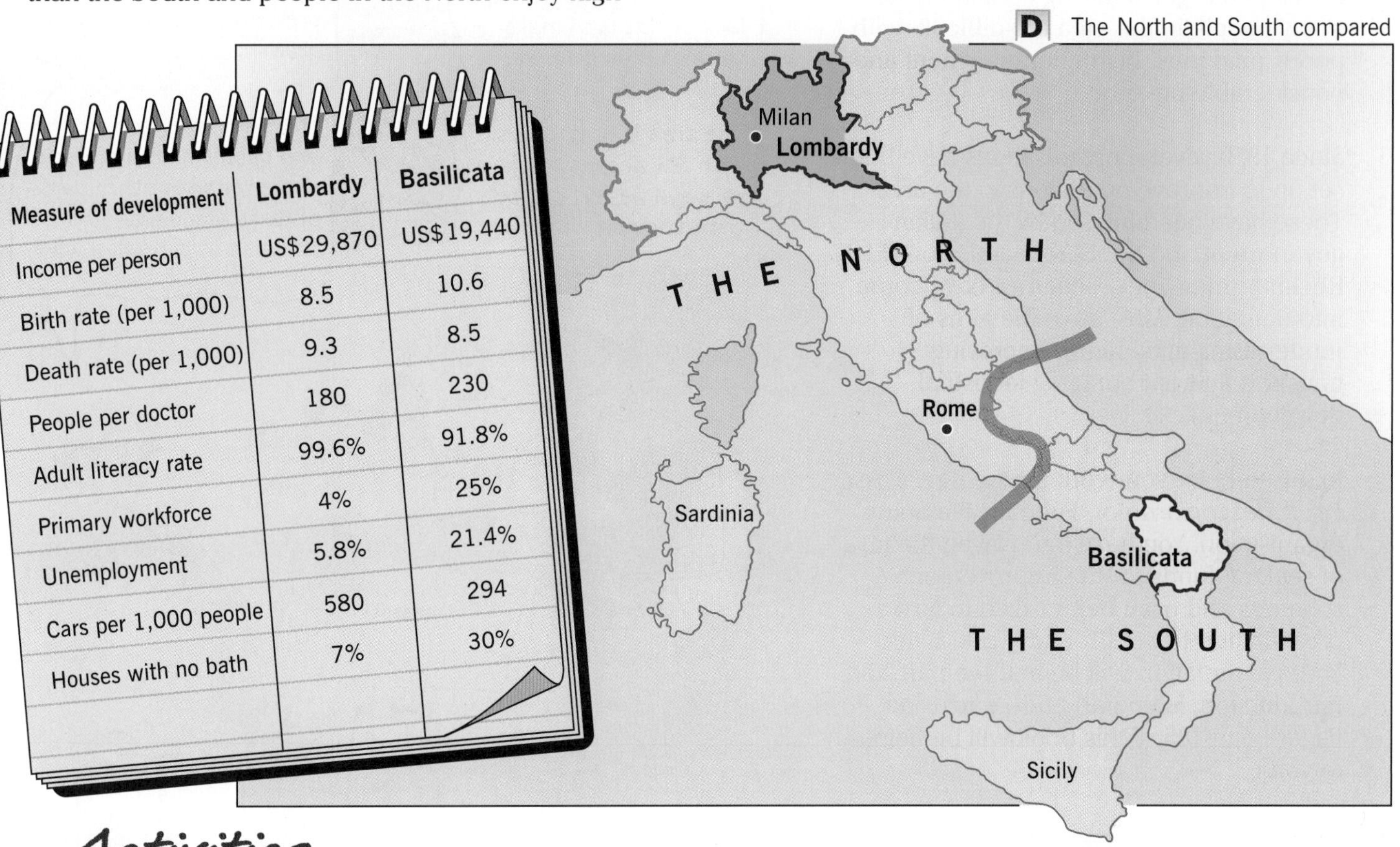

Measure of development	Lombardy	Basilicata
Income per person	US$29,870	US$19,440
Birth rate (per 1,000)	8.5	10.6
Death rate (per 1,000)	9.3	8.5
People per doctor	180	230
Adult literacy rate	99.6%	91.8%
Primary workforce	4%	25%
Unemployment	5.8%	21.4%
Cars per 1,000 people	580	294
Houses with no bath	7%	30%

Milan
Lombardy
THE NORTH
Rome
Sardinia
Basilicata
THE SOUTH
Sicily

Activities

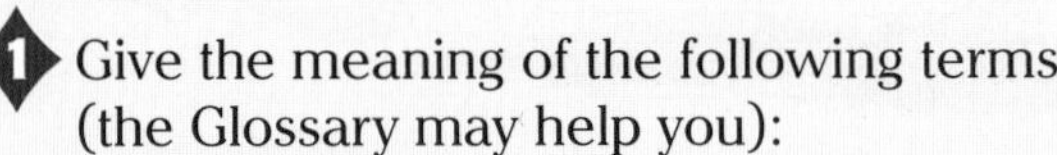

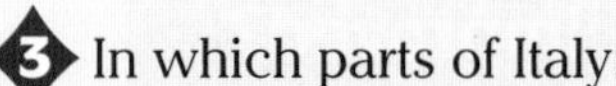

1 Give the meaning of the following terms (the Glossary may help you):

a development
b developed country
c developing country
d standard of living
e quality of life.

2 Work with a partner or in a small group for this activity so that you can discuss ideas and opinions.

a From the drawings in **A**, choose the nine measures that you think are the most important.
b Write out the nine measures in a diamond shape as shown in **E**. Put the most important at the top, the next two below, and so on.
c Give reasons for your choice of first and last.

3 In which parts of Italy:

a are workers likely to earn most money
b are people most likely to be out of work
c is population increasing
d can most people read and write
e is the best housing and medical care?

4 Italy has been described as a modern, wealthy and highly developed nation. Do you agree with this? Give reasons for your answer.

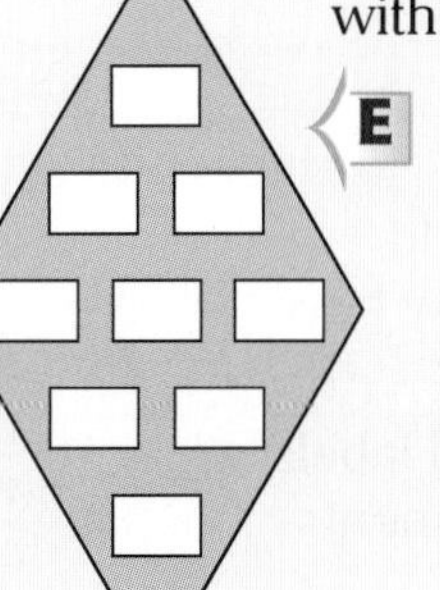

E

Summary

Italy is one of the most developed countries in the world. Development is not spread evenly, though, as some areas in the South have a very poor standard of living.

The Italy enquiry

The South of Italy, or Mezzogiorno as it is called, is one of Europe's poorest and least developed regions. Living conditions for many people in the area are difficult, with poorly paid jobs, high unemployment and considerable poverty.

Since 1950 several organisations have been set up to improve conditions in the area. These have been funded by the Italian government and the European Union (EU). Recently, many new schemes have come into operation. Most have the aims of modernising agriculture, improving transport and encouraging industrial development.

In this enquiry you work for the new *Cassa Per Il Mezzogiorno* or 'Fund for the South' organisation. You have been given the task of seeking funds for the improvement schemes and have been asked to give a presentation to the EU and World Bank. Your presentation will be in three parts: an Introduction, Main part and Conclusion. Pages 96 to 105 of this book will be helpful to you.

How can conditions in the South of Italy be improved?

1 Introduction

a Draw a simple map to show the location of Italy's South. Label the main features.

b Briefly describe the region using information from pages 102 and 103.

c Draw a star diagram to show the main problems of the South.

2 Main part

a Make a simple copy of drawing **B** and add arrowed labels and new features to show how life in the rural areas may be improved.

b Make a copy of map **D** and add arrowed labels and new features to show your development plans for the South.

3 Conclusion

Now write a brief summary of your proposals. Include the following:

a What are the advantages of improving rural conditions?

b Why will some villages have to be abandoned?

c What has happened to the old marshland areas?

d Why is it now possible to have farms on lowland?

e How has transport been improved?

f Why is the industrial estate in a good location?

g What are the advantages of having a seaside resort in this area?

h Which schemes do you think would bring most benefit? Give reasons for your answer.

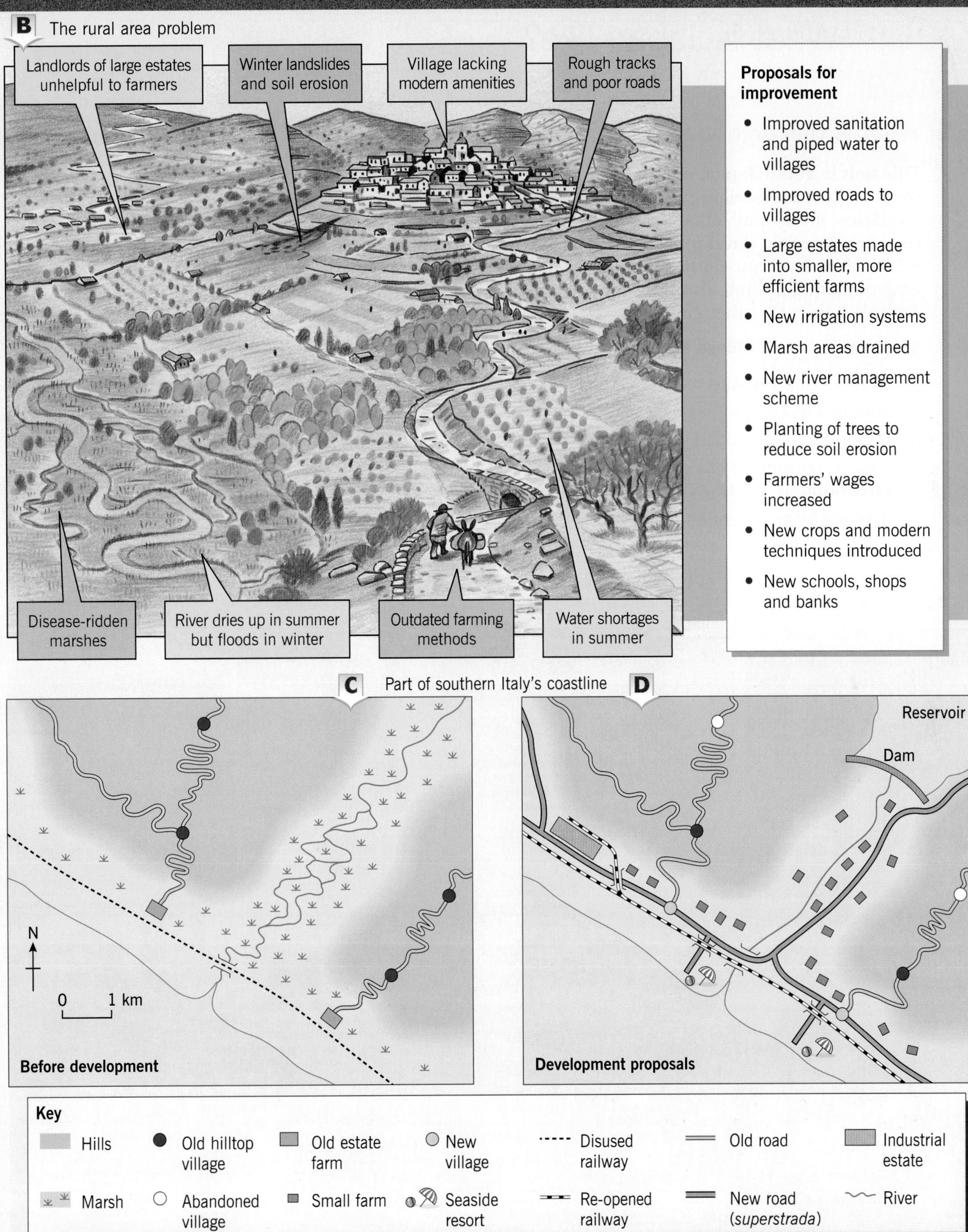

Proposals for improvement

- Improved sanitation and piped water to villages
- Improved roads to villages
- Large estates made into smaller, more efficient farms
- New irrigation systems
- Marsh areas drained
- New river management scheme
- Planting of trees to reduce soil erosion
- Farmers' wages increased
- New crops and modern techniques introduced
- New schools, shops and banks

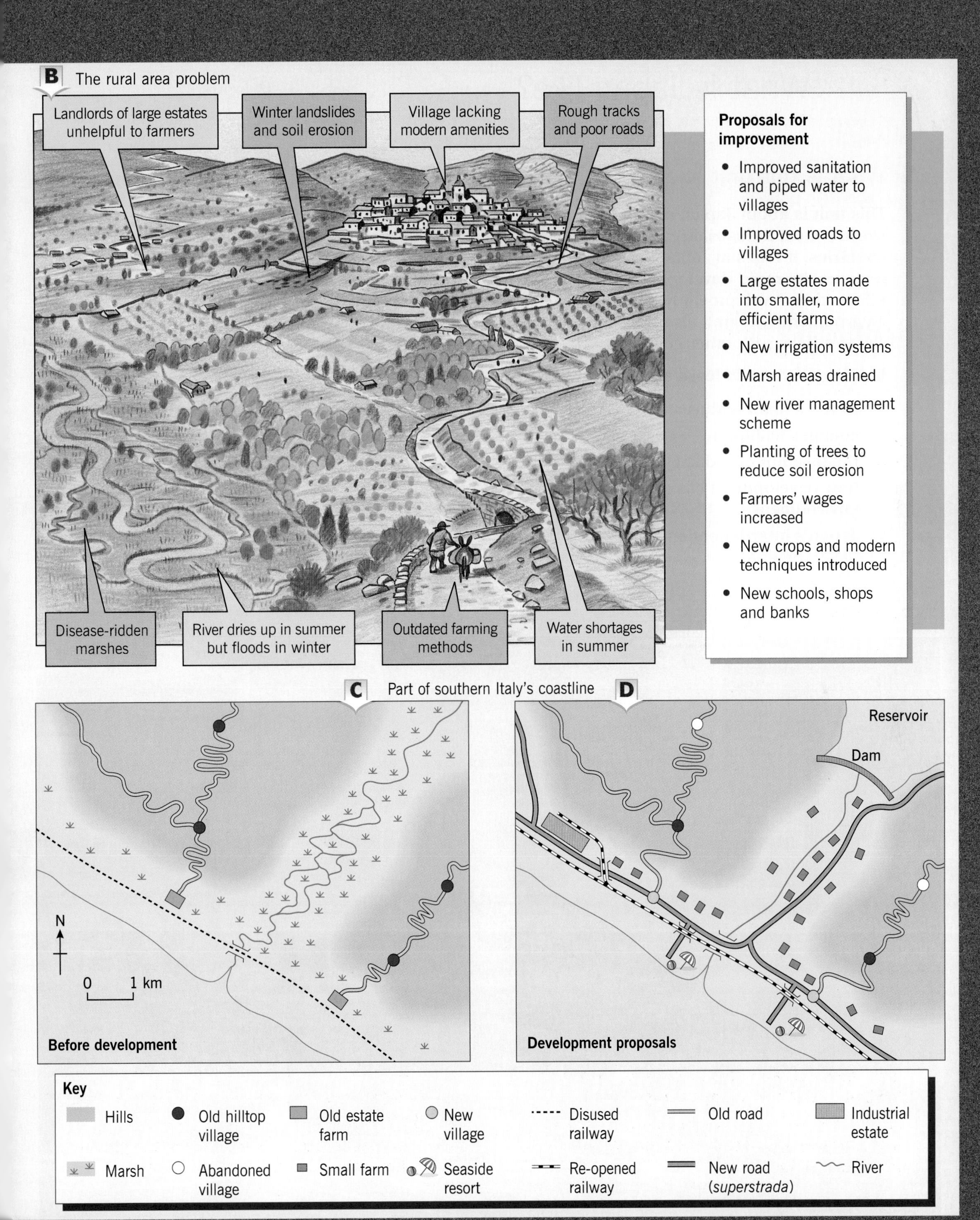

6 Japan, a developed country

What is Japan like?

What is this unit about?

This unit is about Japan, one of the world's most industrialised and richest countries. It looks at Japan's main features, its rapid development and the effects this development has had on the environment. The unit also looks at Japan's links with other countries.

In this unit you will learn about:

- **Japan's main features**
- **changes in Japan**
- **the location of industry**
- **how development has affected the environment**
- **Japan's interdependence and development.**

Why is learning about Japan important?

Learning about Japan will give you an interest and knowledge of people and places that are very different from those found in the UK. It will also help you learn about a country which makes many of our electronic goods and plays an important part in the world's trading system.

The unit can also help you to:

- broaden your knowledge of the world
- learn about different landscapes, natural hazards and climate
- understand ways of life that are different from your own
- develop an interest in other countries.

A Tokyo

B High-rise flats in Tokyo

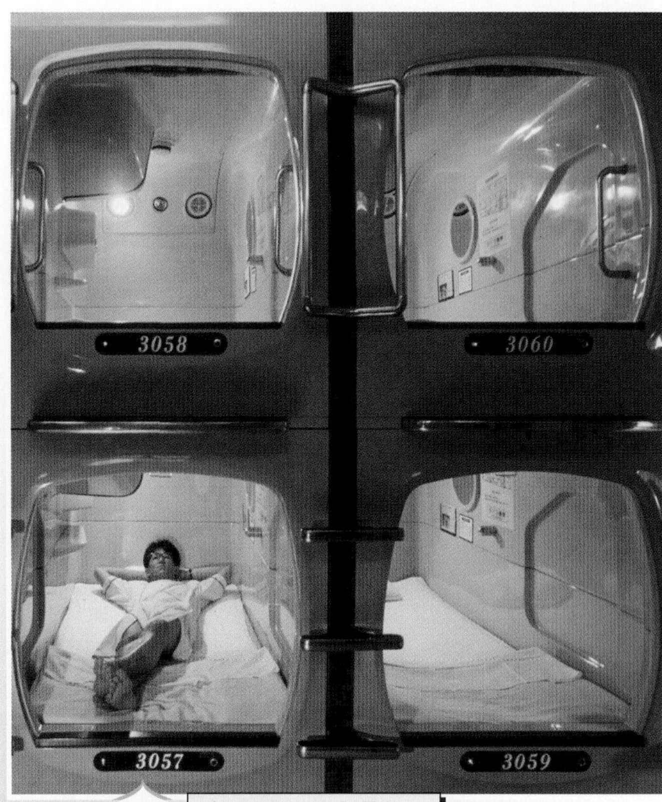

C Capsule hotel, Tokyo

D Mount Fuji from Tokyo

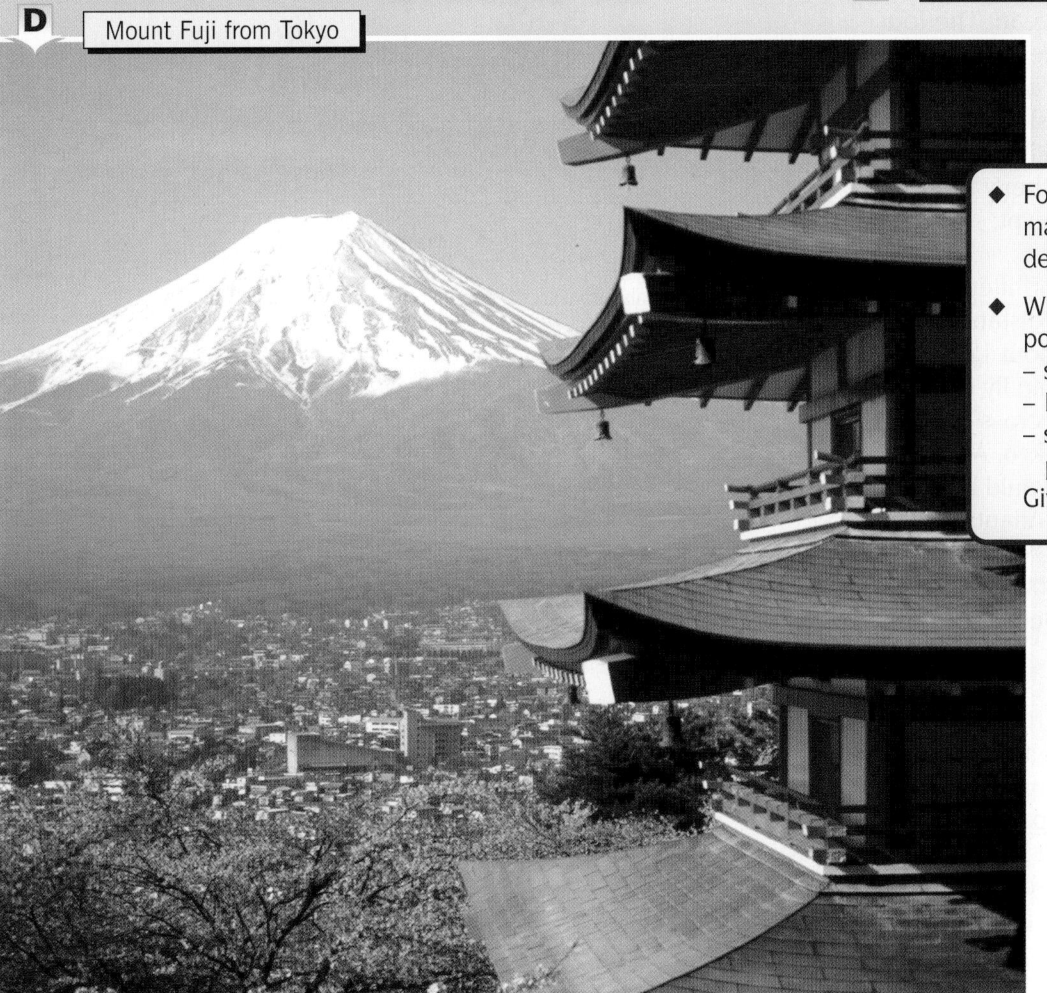

- For each of the four photos, make a list of words to describe it.
- What are the good and bad points of:
  - shopping in Tokyo in photo **A**
  - living in the flats in photo **B**
  - staying at the hotel in photo **C**?

  Give reasons for your answers.

Where is Japan?

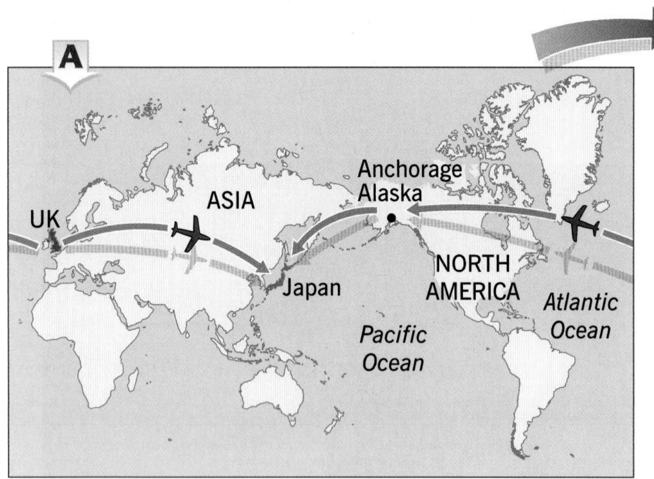

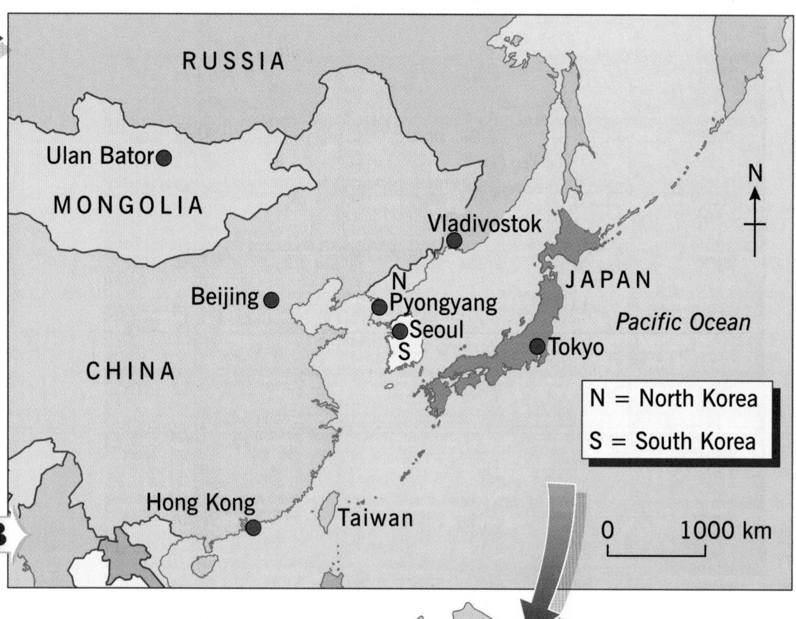

Japan's nearest neighbours **B**

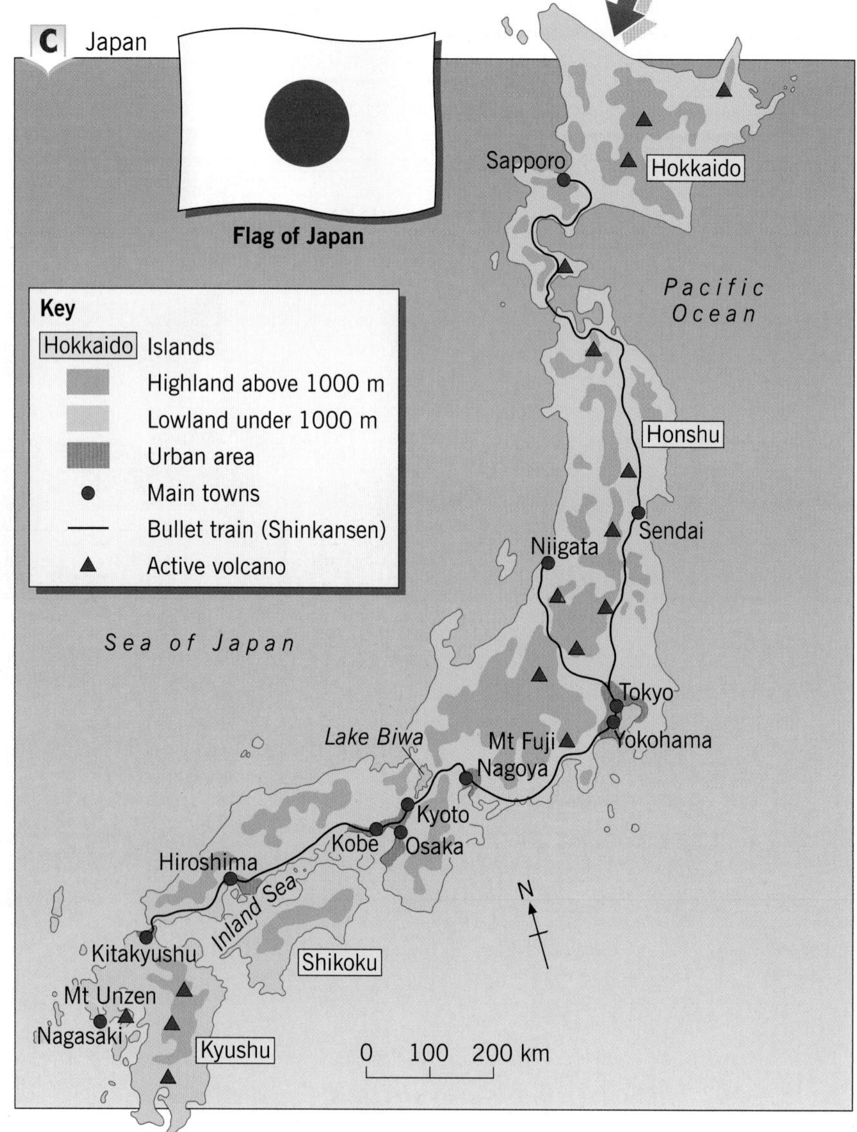

C Japan

Japan is located about 200 km off the coast of Asia in the Pacific Ocean. The four main islands that make up the country stretch from Hokkaido in the north to Kyushu in the south. The most northerly point is on about the same latitude as Venice in Italy. The most southerly point is about the same as Cairo in Egypt.

Japan is a long way from the United Kingdom. The shortest flying distance is 10,320 km and would take about 12 hours. On this route the aircraft would fly east from the UK over Russia and cross the Sea of Japan before landing in Tokyo. An alternative but longer route would be to fly west from the UK across the Atlantic Ocean and North America to Alaska for refuelling. The journey would then continue across the Pacific Ocean and eventually on to Japan.

Japan's nearest neighbours are to the north and west. Russia is just 40 km from the northern tip of Hokkaido, whilst a short distance across the Sea of Japan are North and South Korea. The nearest land to the east is California in the USA, some 8,800 km and ten hours' flying time across the Pacific Ocean.

D A winter snow scene in Hokkaido

E Palm tree and beach on Kyushu

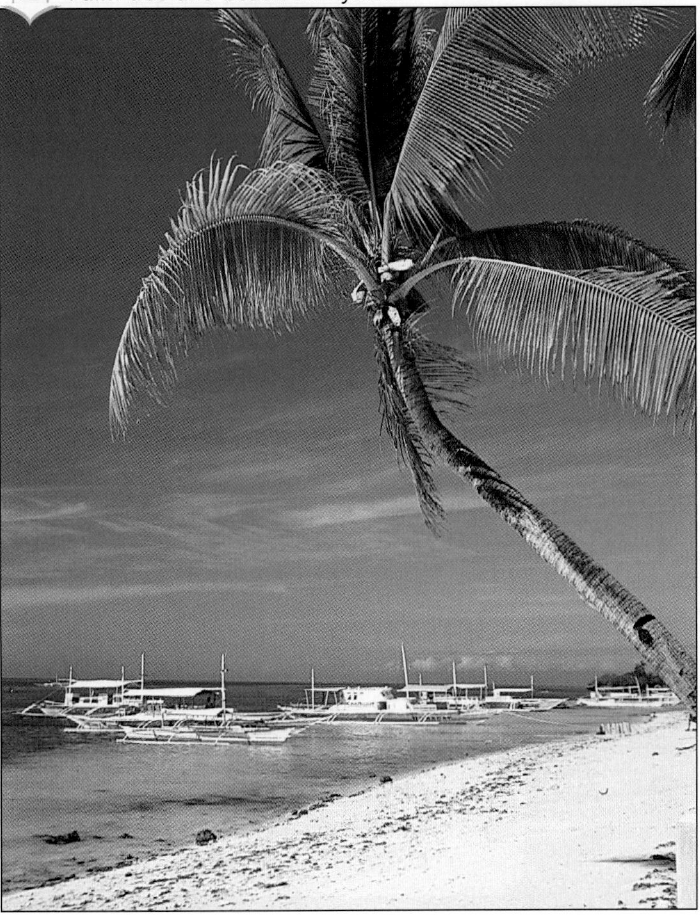

Activities

1 Copy and complete quizword **F**, using the following clues:

- **a** Country to the east of Japan
- **b** Country that has Beijing as its capital
- **c** Separates the USA from Japan
- **d** Country made up of four main islands
- **e** Large city north-east of Mount Fuji
- **f** Japan's nearest neighbour
- **g** Two countries share this name.

Make up a clue for downword **h**.

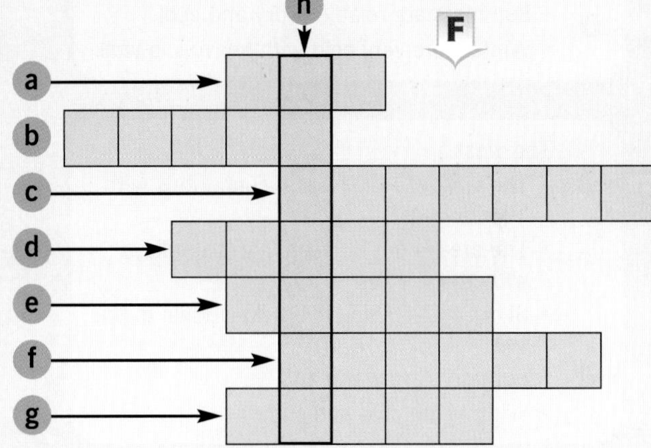

2 Look at map **C**.

- **a** Name the towns on Japan's Inland Sea.
- **b** Which island has no large towns?
- **c** Name two active volcanoes.
- **d** What is the distance by bullet train:
  - across Japan from Tokyo to Niigata, and
  - the length of Japan from Kitakyushu to Sapporo?

3 Imagine that you are on holiday in Japan and have just visited Hokkaido and Kyushu. Send a postcard home describing each of the places. Use photos **D** and **E** and include the words below.

Japan – Kyushu – Hokkaido – north – south – hot – cold – sunny – cloudy – dry – snowy – mountains – beaches – skiing – warm sea – tropical vegetation

Summary

Japan lies off the east coast of Asia in the Pacific Ocean. Its nearest neighbours are Russia, North Korea, South Korea and China.

What are Japan's main physical features?

This picture of Japan is based upon photos taken from 830 kilometres above the earth's surface by an orbiting satellite. The colours and relief have been changed slightly (enhanced) to make the features clearer and easier to identify. Notice how mountainous the country is and how little flat land can be seen.

The north

- Hokkaido is remote and mountainous.
- Snow-capped peaks cover the centre of the island.
- Summers are relatively dry and cool.
- Winters are very cold with heavy snowfall.

The west

- The west of Honshu is mountainous with little flat land.
- The area is cut by deep, narrow valleys with steep sides.
- Short, fast-flowing rivers flow down to the sea.
- Summers are warm and wet.
- Winters are cold and snowy.

The south

- Kyushu has many active volcanoes, crater lakes and hot springs.
- Coral reefs may be found in the warm seas along the coast.
- The climate is almost tropical with hot, wet summers and warm sunny winters. Typhoons are common in autumn.
- The island has tropical plants and a lush vegetation.

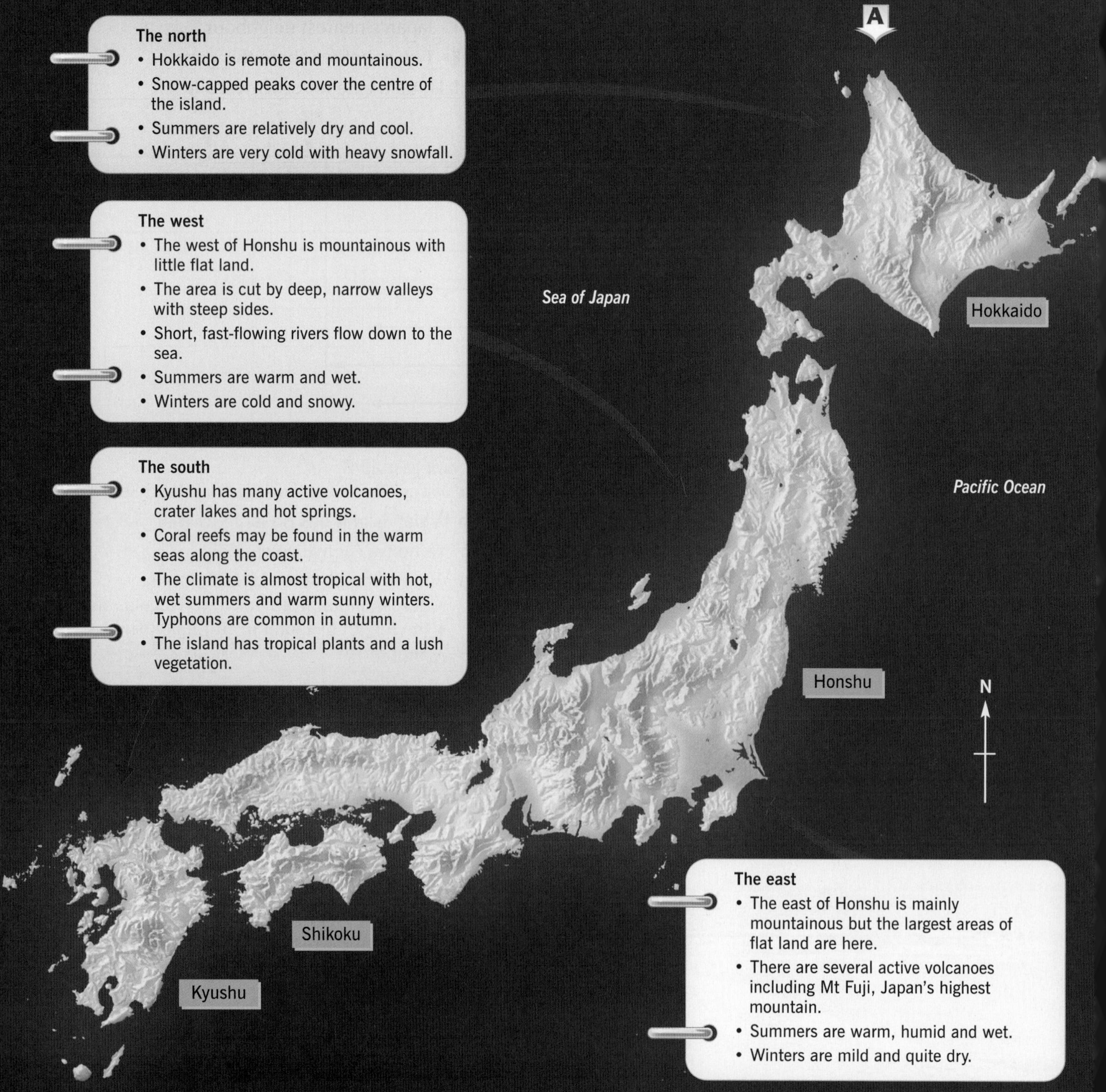

A
Sea of Japan
Hokkaido
Pacific Ocean
Honshu
N
Shikoku
Kyushu

The east

- The east of Honshu is mainly mountainous but the largest areas of flat land are here.
- There are several active volcanoes including Mt Fuji, Japan's highest mountain.
- Summers are warm, humid and wet.
- Winters are mild and quite dry.

Japan is a land of contrasts with great variations in landscape and climate. The country itself consists of four large islands and over 1,000 smaller ones. The four largest are Honshu – which is a little larger than Great Britain – Hokkaido, Kyushu and Shikoku. The islands were formed by volcanoes, many of which are still active. As map **A** shows, most of Japan is mountainous with less than 20 per cent of the country being flat enough for farming or settlement.

The climate of Japan varies considerably from place to place. There are two main reasons for this. The first is that Japan is a very long country and stretches some 3,000 km from north to south. This means that the south is much warmer than the north. The second reason is that Japan is a mountainous country. As temperatures usually decrease with height, the highland areas can be very much colder than the coastal lowlands. They are also wetter and receive heavy snowfall in winter.

Japan can be a dangerous place in which to live. It suffers from several natural hazards, some of which are shown in drawing **B**. Some of the hazards are due to Japan's location on an active plate boundary where earthquakes and volcanic eruptions are common (see maps **A** and **B** on pages 28 and 29). Others are linked to extremes of climate.

B

Japan's natural hazards

- **Earthquakes** are the most serious threat, with over 7,000 each year. Some are severe but most are just gentle tremors.
- Japan has 70 **active volcanoes**, including Mount Fuji.
- A **tsunami** is a giant wave, or series of waves, usually caused by an earthquake or volcanic eruption on the sea bed.
- **Heavy winter snowfall** is common in the north and in the mountains.
- **Typhoons** are tropical storms with winds up to 200 km/hr and torrential rain.

Yet despite these hazards, the Japanese believe that nature has been kind to them. They say that:

- the heavy rainfall has covered the mountains with forest and provides water for irrigation
- the warm, wet summers are ideal for growing rice
- volcanic eruptions give fertile soil
- the fast-flowing rivers supply fresh water and produce hydro-electricity
- the warm seas are an important source of food.

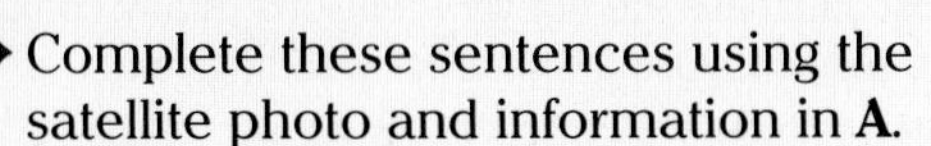

1 Complete these sentences using the satellite photo and information in **A**.

a The four main islands in order of size are ...
b The most northerly island is ...
c The most southerly island is ...
d The warmest region is ...
e The region with most snowfall is ...
f The island most affected by tropical storms is ...
g The region with most flat land is ...
h The two islands with least flat land are ...

2 Explain why earthquakes and volcanic eruptions are common in Japan. Page 30 will help you.

C

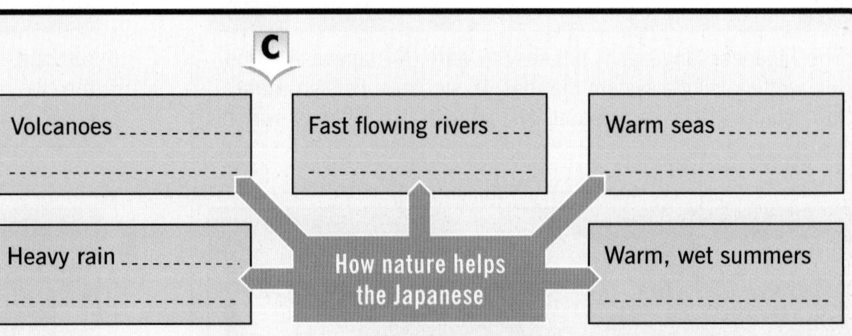

3 Copy diagram **C** and add a sentence to each box to show how the Japanese believe nature has been kind to them.

Summary

Japan is a mountainous country with little flat land. Natural hazards caused by earth movements and extremes of climate cause problems for people.

How has Japan changed?

Japan is a country steeped in tradition yet with a modern outlook on life. The reasons for this go back many years. For over 200 years Japan chose to be isolated from the rest of the world and until 1897 no foreigners were allowed into the country. A series of disastrous wars then hindered development and slowed progress. Recovery finally began in the 1950s, and by 1968 Japan had become the world's second most powerful economy. This period of fast growth is often called the '**economic miracle**'.

Traditional way of life

Japan's isolation enabled it to develop a unique culture and way of life. As this page shows, Japanese traditions are very different from those of other countries.

C

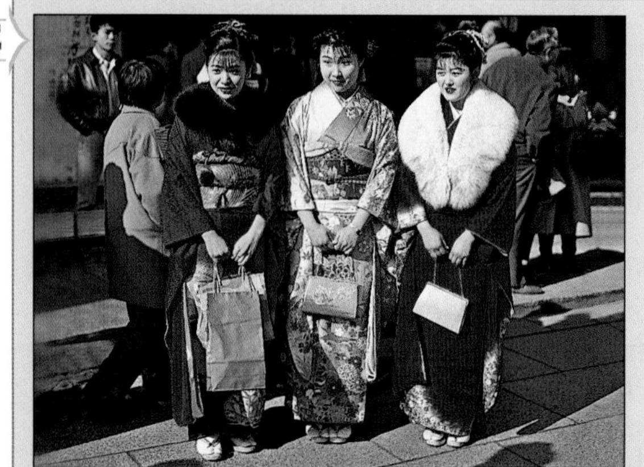

The traditional dress is a long, loose-fitting robe (*kimono*), a waist sash (*obi*) and flat shoes (*zori*). Shoes are always removed before entering a house.

A

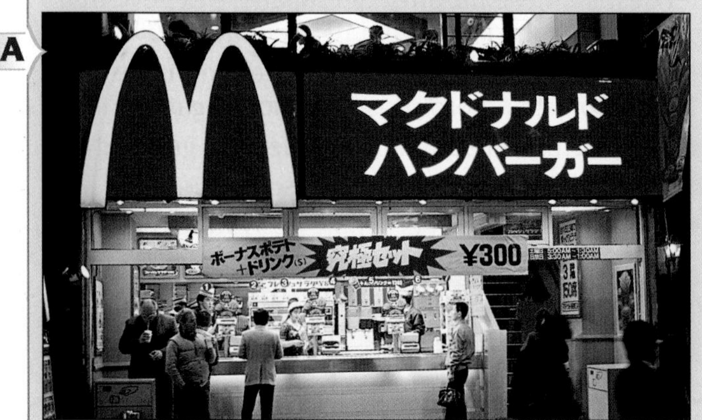

The Japanese language is not easy to learn. All Japanese words are written using a system that has 46 symbols. Foreign words are usually written in a second system using 46 different symbols.

D

Although most Japanese are not very religious, many combine the two main religions, using Shinto for weddings and Buddhism for funerals.

B

Houses are small, with sliding paper partitions. People sit on cushions on the floor and bedding is stored away during the day.

E

Several Japanese customs have become well known worldwide. These include sumo wrestling, kendo fencing, karate, paper folding (*origami*), and miniature plant growing (*bonsai*).

Modern Japan

The very rapid growth of cities and industry in the last 50 years has dramatically altered the Japanese way of life and made Japan rich. Although Japan has less than 3 per cent of the world's population it earns nearly 10 per cent of the world's money.

Most houses are small and are built very close together. They are full of modern gadgets yet they keep the characteristics of traditional homes.

City centres have huge shopping centres with department stores full of high-quality goods, mostly made in Japan. Tall office blocks tower above congested roads. Streets are full of workers and shoppers during the day. In the evening people seek the active nightlife. Most people wear western-style dress.

Most Japanese spend long hours at work and have few holidays. They are extremely loyal to their family and place of work. Politeness is basic to their way of life. They are patriotic, highly educated and skilled.

Many children go to extra classes at night because exam failure is considered a family disgrace. Over 90 per cent of children over the age of 16 stay on at school and one-third go on to university.

F Tokyo at night

Activities

1 **a** What does 'isolated' mean?

b How did isolation help Japan develop its own traditions?

c Why was Japan's progress so slow at first?

d What is meant by the term 'economic miracle'?

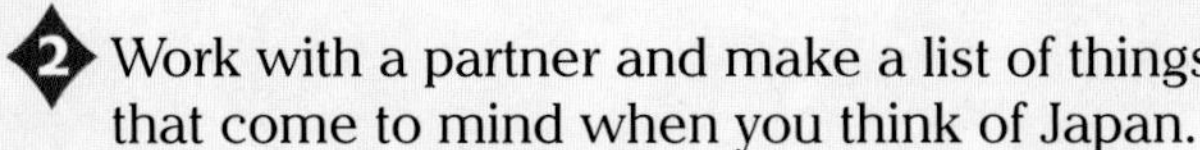

2 Work with a partner and make a list of things that come to mind when you think of Japan.

3 Complete a fact file comparing traditional and modern Japan. Use the headings shown in drawing **G**. Write a sentence for each of the points.

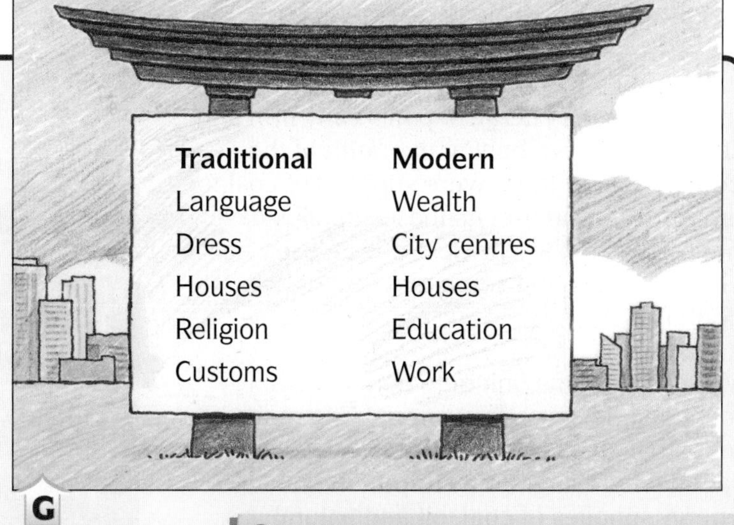

G

Summary

Living on a group of islands, the Japanese have developed a distinctive culture and way of life which combines tradition with modern living.

Where is Japan's industry located?

The growth in industry

In 1945 Japan had hardly any industry. By 1990 it had become the most industrialised country in the world. How did it achieve this success considering that it lacked the basic industrial needs of flat land, raw materials and sources of energy?

It began by using its limited amounts of iron ore and coal to make steel. The steel was used to build large ships, new factories and houses. The ships were designed to carry to Japan those raw materials which the country needed. These included iron ore, coking coal and aluminium for its industry, and coal and oil to provide its energy.

Japan then turned its attention to producing cars and to developing electronics and high-technology industries. The Japanese have become the world leaders in producing video recorders and camcorders, stereo sound and compact disc systems, cameras and computer parts. The money which Japan makes by selling these goods easily pays for the cost of buying the raw materials and foodstuffs.

The location of industry – sources of energy

Industry needs energy. Japan's early iron and steel works were built in the north of the country where there were supplies of coal for energy, and iron ore. As these supplies have been used up then present-day industry has chosen sites in the south and east of the country.

Map **A** shows the link between industrial location and sources of energy. The main industrial areas lie along the Pacific coast of Japan where large ports import the needed energy supplies of coal, oil and natural gas. There is a smaller industrial area along the west coast where there are many nuclear and hydro-electric power stations.

A Location of Japanese industry

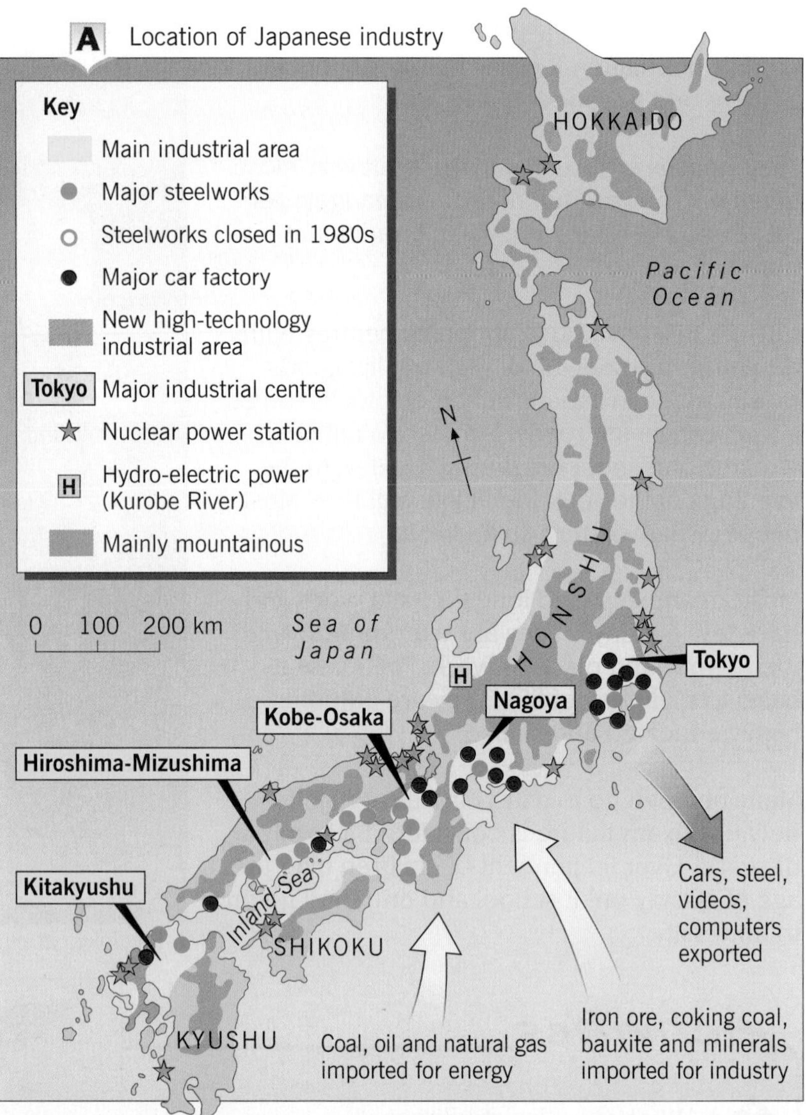

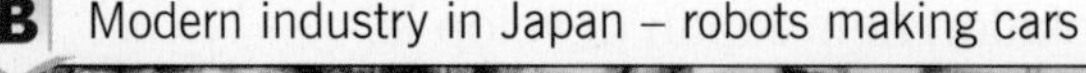

B Modern industry in Japan – robots making cars

Other location factors

Diagram **C** gives other reasons why Japan's main industrial regions should be located in the south and east of the country. The Japanese themselves consider the two most important things in finding the **site** for a new industry to be the distribution of lowland and people, and the location of sheltered deepwater ports. There is now such a shortage of building space that land for any new large factory has to be reclaimed from the sea (photo **D**).

C

D Mazda car factory, Hiroshima

Activities

1 Write out the sentences below in the correct order to describe the growth of industry in Japan.

- Ships and factories were built using steel.
- Car and electrical industries developed.
- Steel was first made from local raw materials.
- Money earned from selling goods was spent on buying raw materials.
- Ships were used to import raw materials.

2 Make a simple sketch of the car factory shown in photo **D**. Add the following labels to show why it was a good site to locate a large modern car factory.

- Large local market for cars
- Good accessibility by motorway
- Flat land reclaimed from sea
- Deep and sheltered harbour
- Port to import raw materials and energy
- Port to export finished cars
- Attractive location with hills and coast

Summary

Although some industry in Japan is built near to sources of energy, the most important location factors are the distribution of lowland and people, and the position of sheltered deepwater ports.

Sustainable development in Japan

We have seen that Japanese industry grew rapidly after 1945. This development was accompanied by a rapid growth in population and an expansion of cities.

Diagram **A** shows that most of Japan is mountainous and forested and that only 17 per cent is suitable for living and working in. This led to tremendous competition for space for housing, industry, recreation and transport. This competition led, in turn, to the creation of many environmental problems, especially around Tokyo Bay.

Diagram **B** describes some of the environmental problems created by rapid urban growth and economic development in the Tokyo area. The photo is a satellite image of the same area. On it flat land is coloured green. Notice how little there is of it.

17% maximum for living and working
Housing, industry and transport 3%
Farmland 14%
Mountains and lakes 16%
Forest 67%

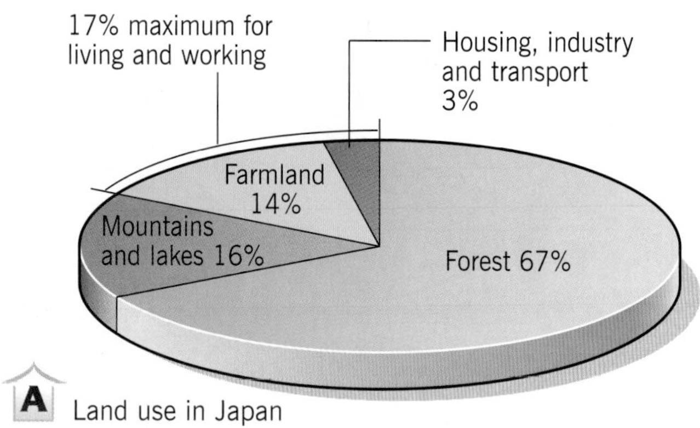

A Land use in Japan

B Causes of pollution in the 1980s

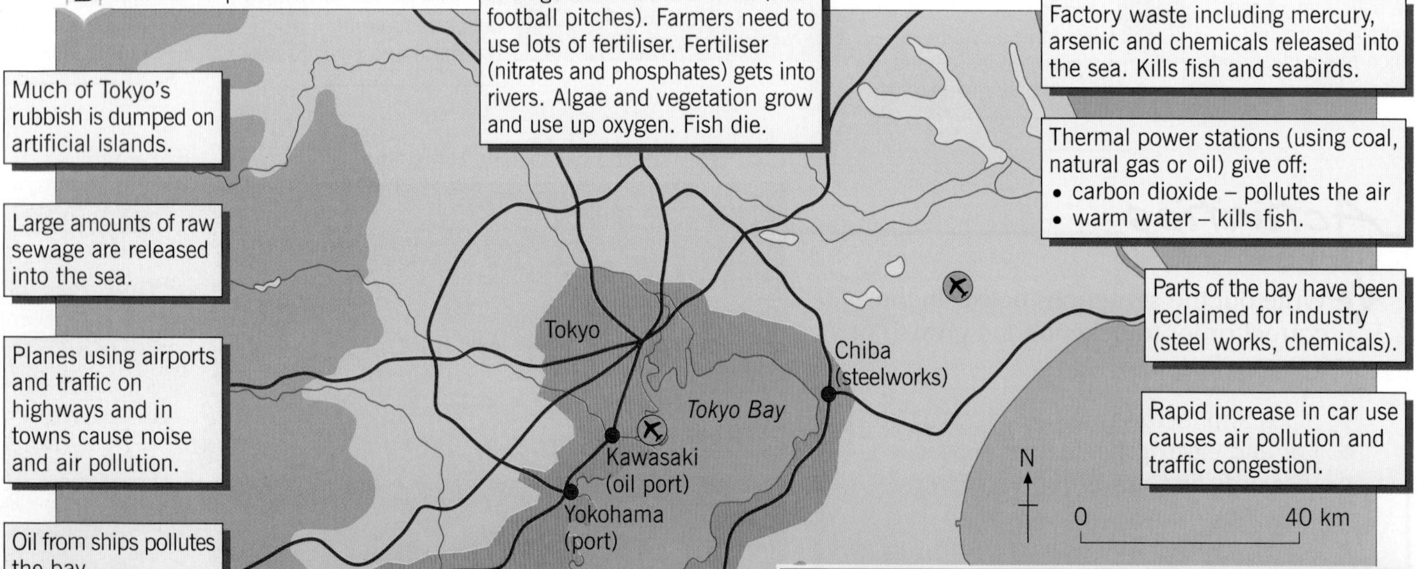

Much of Tokyo's rubbish is dumped on artificial islands.
Average farm size is 1.1 ha (two football pitches). Farmers need to use lots of fertiliser. Fertiliser (nitrates and phosphates) gets into rivers. Algae and vegetation grow and use up oxygen. Fish die.
Factory waste including mercury, arsenic and chemicals released into the sea. Kills fish and seabirds.
Thermal power stations (using coal, natural gas or oil) give off:
• carbon dioxide – pollutes the air
• warm water – kills fish.
Large amounts of raw sewage are released into the sea.
Parts of the bay have been reclaimed for industry (steel works, chemicals).
Planes using airports and traffic on highways and in towns cause noise and air pollution.
Rapid increase in car use causes air pollution and traffic congestion.
Oil from ships pollutes the bay.
Loss of wildlife habitats.
Tokyo
Chiba (steelworks)
Tokyo Bay
Kawasaki (oil port)
Yokohama (port)
N
0
40 km
Key
Airport
Built-up area
Highland
Lowland
Highway

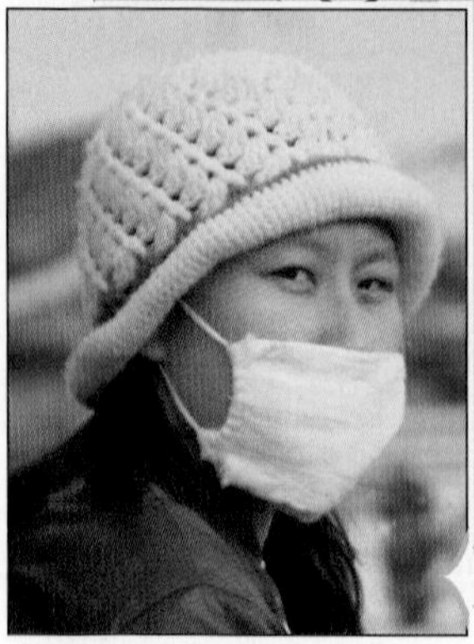

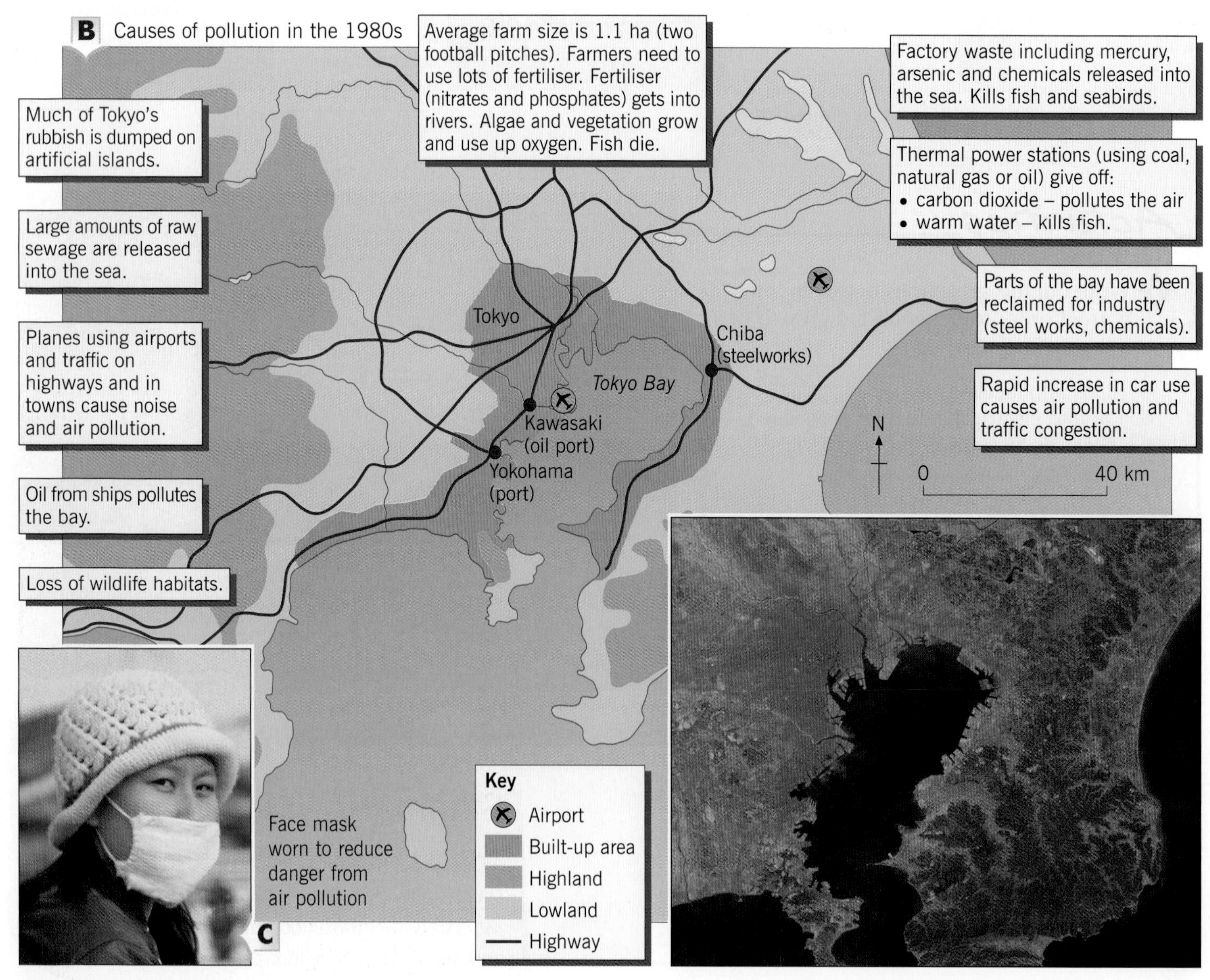

C Face mask worn to reduce danger from air pollution

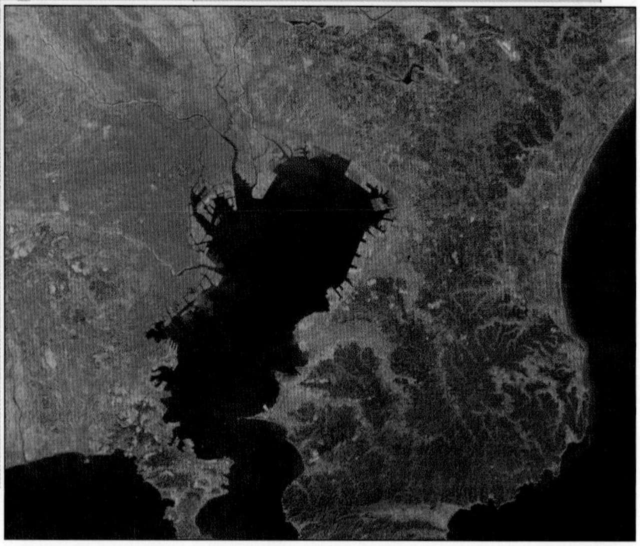

The environment today

As Japan developed its industries and its people increased in wealth, little thought was given as to how this would affect the environment. Water supplies and the air became heavily polluted. This created serious health problems for people, threatened wildlife and destroyed the attractive scenery.

By the early 1980s, the Japanese realised the need for a more **sustainable development**. Sustainable development is a way of improving people's standard of living and quality of life without wasting resources or harming the environment in which they live. The Japanese decided they needed to use some of their wealth and technology to reduce environmental pollution, even if this meant they became rich less quickly.

Diagram **D** shows some of the ways by which levels of air, water and noise pollution have been either reduced or controlled. Even so, pollution created in earlier years cannot be cleaned up overnight.

D

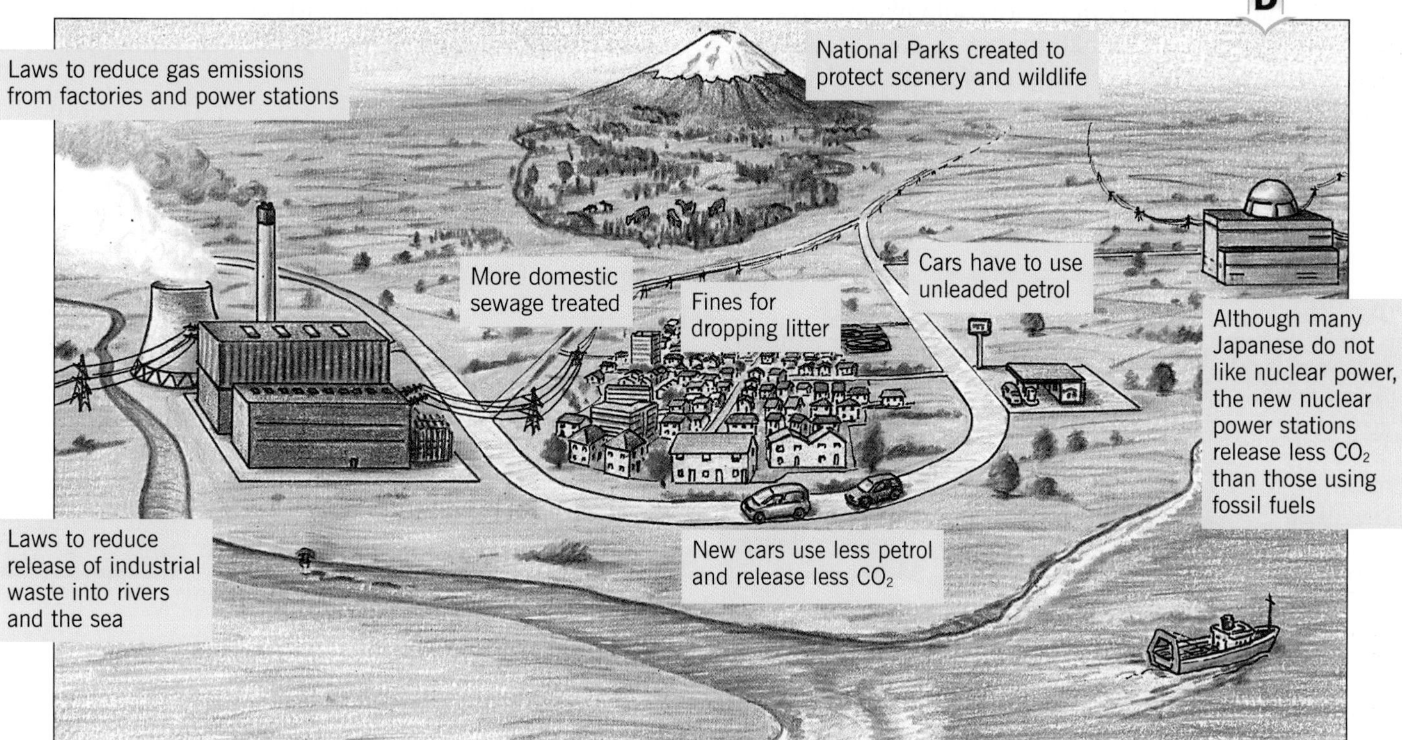

Activities

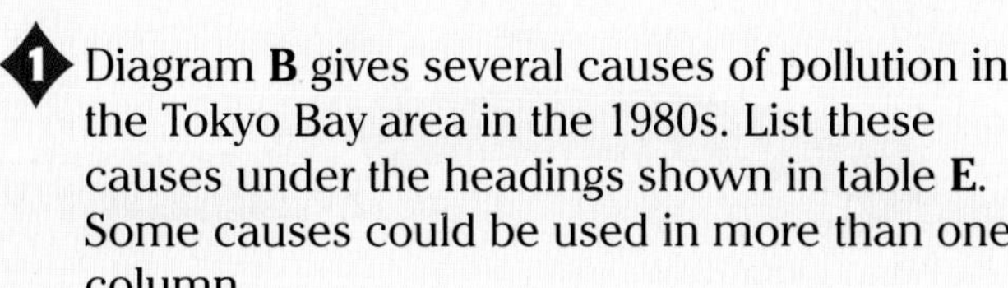

1. Diagram **B** gives several causes of pollution in the Tokyo Bay area in the 1980s. List these causes under the headings shown in table **E**. Some causes could be used in more than one column.

E

Water	Air	Noise	Visual

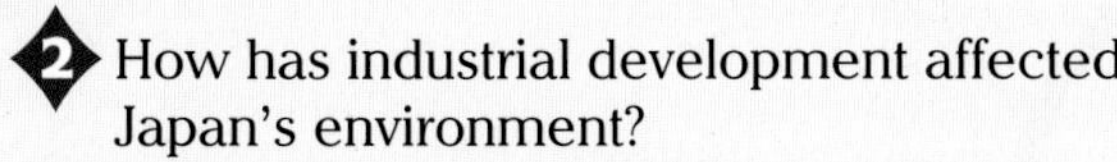

2. How has industrial development affected Japan's environment?

3. **a** What is sustainable development?
   b Why did Japan become interested in sustainable development?
   c Describe the attempts made to try to reduce both water and air pollution.
   d Describe the attempts made to try to protect both resources and scenery and wildlife.

Summary

The growth of industry and cities caused widespread pollution in Japan. By adopting a more sustainable development it is hoped that resources will be conserved, the environment will be protected and the standard of living will still improve.

How interdependent is Japan?

When countries work together and rely on each other for help they are said to be **interdependent**. Being interdependent can help a country progress and improve its standard of living.

One of the main ways that countries become interdependent is by selling goods to each other. They buy things that they need or would like to have. They then sell things to make money to pay for what they have bought. The exchanging of goods and materials like this is called **trade**. Goods sold to other countries are called **exports**. Goods that are bought by a country are called **imports**.

Timeline **A** shows how Japan has become one of the world's leading trading nations. It is responsible for 9 per cent of the world's exports and 6 per cent of the world's imports. Japan needs to trade with other countries mainly because it has very few **natural resources** of its own. **Raw materials** like oil, coal and iron ore need to be imported so that Japanese factories can produce goods to be sold at home and abroad. With limited land available for farming, food is another major import.

Although map **B** shows that the USA is Japan's most important trading partner, imports and exports to and from nearby countries have increased in recent years. Today, over 50 per cent of Japan's trade is with its Asian neighbours like China, South Korea and Indonesia.

A

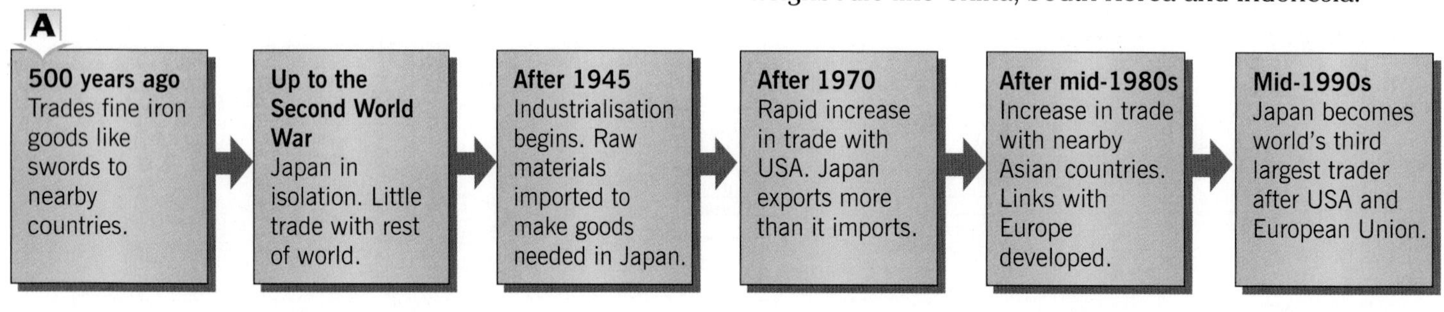

B Japan's main trading partners

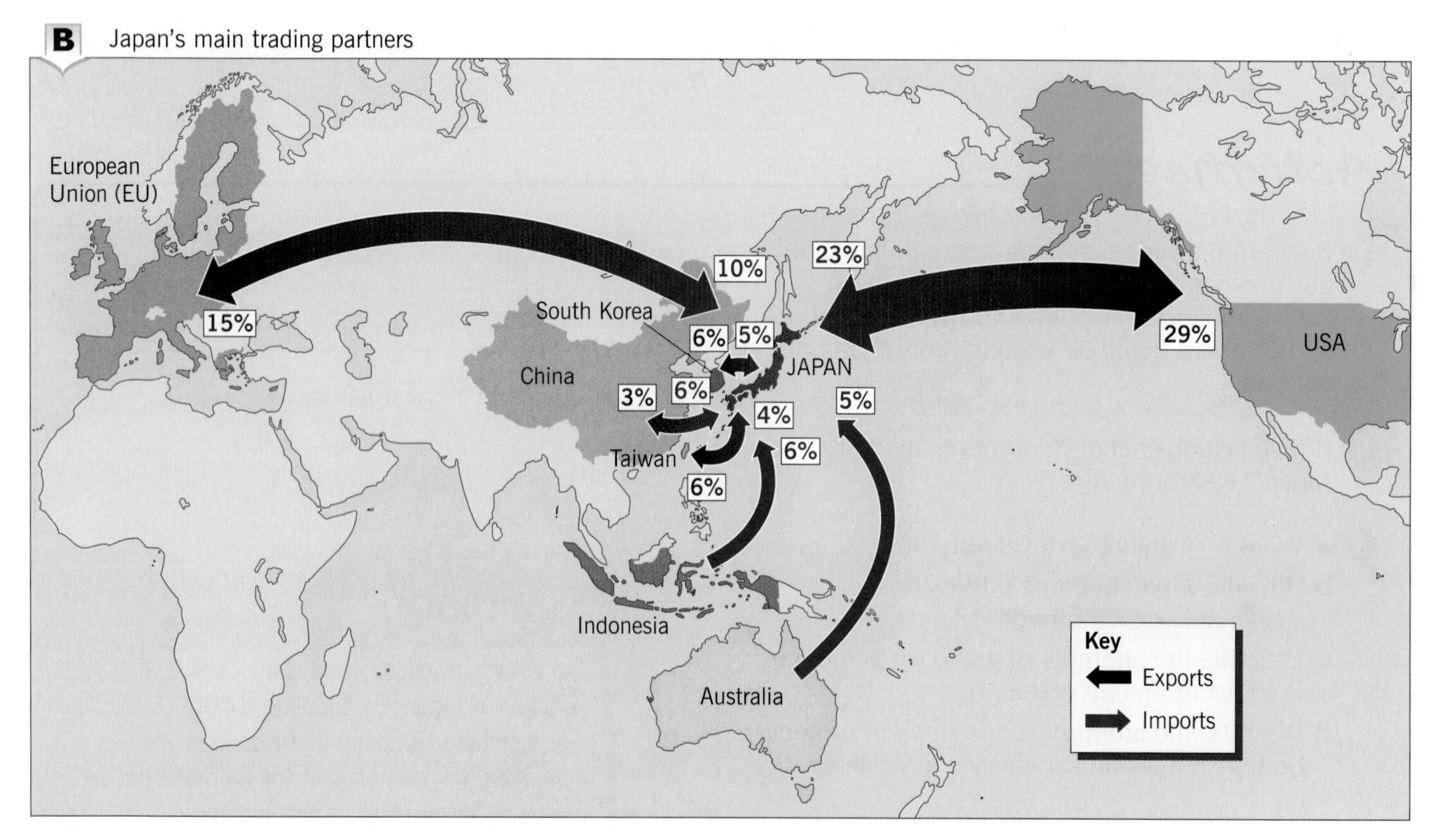

European Union (EU)
15%
10%
South Korea
6%
5%
China
3%
6%
JAPAN
4%
Taiwan
6%
6%
5%
23%
29%
USA
Indonesia
Australia

Key
Exports
Imports

The difference between imports and exports is the **balance of trade**. As drawing **C** shows, Japan earns more from its exports than it spends on its imports. This is something that very few countries achieve and is called a **trade surplus**. Japan's trade surplus is typical of many of the world's more developed countries. It imports relatively low-value foodstuffs and minerals but exports higher-value **manufactured goods**.

Japan's trade surplus is one of the biggest in the world and has brought great wealth to the country. Some of this wealth is now being used overseas, as shown in drawing **D**. This has further increased Japan's interdependence with other countries.

C

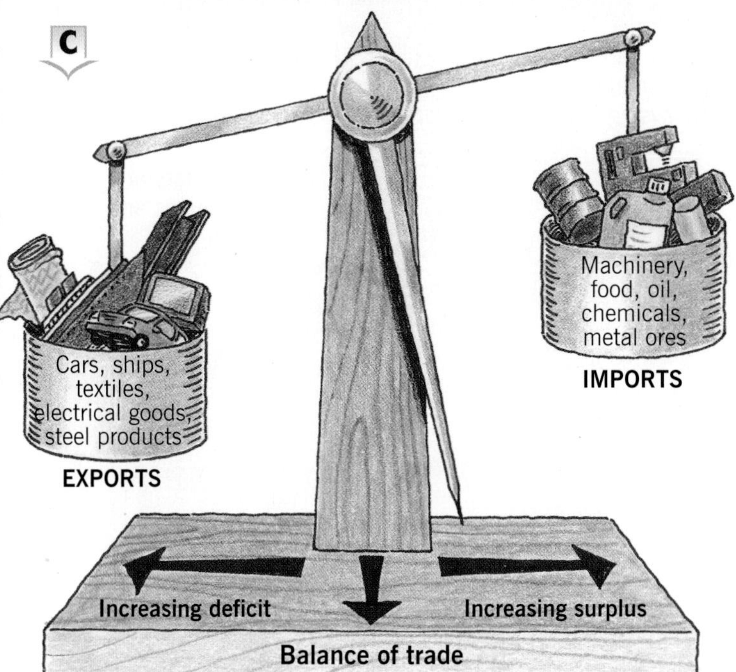

D Japan's interdependence

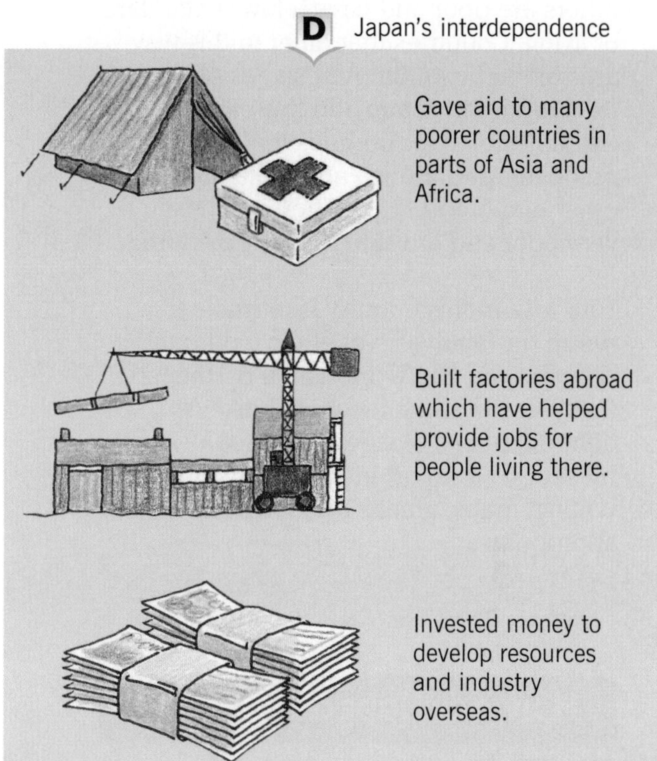

Activities

1. Give the meaning of each of the following terms. You may need to use the Glossary.
   - imports
   - exports
   - trade
   - natural resources
   - raw materials
   - manufactured goods
   - trade surplus
   - trade deficit

2. From map **B**, list Japan's export and import markets in order of size. In each case give the largest first.

3. Make a larger copy of diagram **E** and add the following information to the boxes in the correct order.
   - that are sold overseas
   - to make high-value products
   - to use in factories
   - low-value raw materials
   - which makes the country rich

E

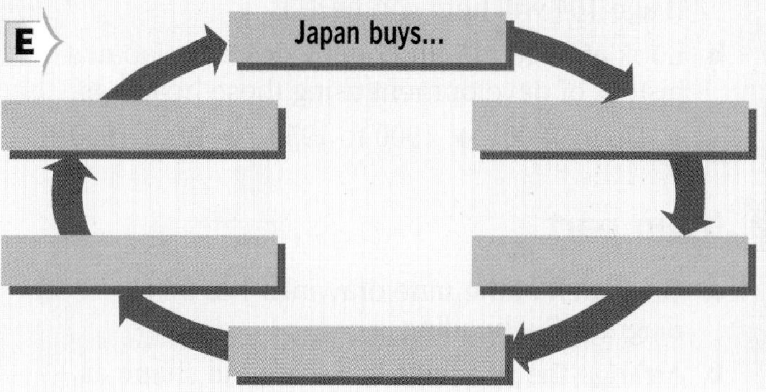

4. **a** What is meant when a country is said to be 'interdependent'?
   b Give four examples of Japan's interdependence with other countries. For each example explain the likely benefits for Japan and the other country.

Summary

The need to import raw materials and foodstuffs and to export manufactured goods has made Japan increasingly interdependent.

The Japan enquiry

All countries are different. Some are rich and have a high **standard of living** whilst others are poor and have a lower standard of living. Countries that differ in this way are said to be at different stages of **development**. Japan and the UK are examples of rich countries that are said to be **developed**. Kenya and India are at the other end of the scale. They are mostly very poor and are said to be **developing**.

Your task in this enquiry is to make a report on Japan's level of development. It is best if you can work with a partner so that you can share ideas and discuss different points of view. You should present your report in three parts and use writing, maps, graphs and diagrams where appropriate.

How developed is Japan?

1 Introduction

a What is meant by the term 'development'? (Page 104 will help you here.)

b Look at page 114 and briefly describe Japan's history of development using these headings:

- Up to 1900
- 1900 to 1950
- After 1950

2 Main part

a Give each of the nine drawings 1 to 9 in diagram **C** a heading.

b Arrange the headings in a diamond shape as shown in diagram **B**. Put the heading that best shows Japan to be developed at the top, the next two below, and so on.

c Draw three bar graphs to show the information in drawings 10, 11 and 12 in diagram **C**. Arrange the bars as follows:

- wealth – highest income on the left
- education – most able to read and write on the left
- food – highest-quality food supply on the left.

3 Conclusion

a In terms of wealth and other economic factors, what evidence is there to suggest that Japan is one of the world's most developed countries?

b How developed is Japan in terms of social and cultural factors? (Pages 114 and 115 may also help you here.)

c Some people say that the Japanese have 'sacrificed their quality of life in order to improve their standard of living'. Do you agree with this? Give reasons for your answer.

Japan – measures of development

C

1

2

3

4

5

6

7

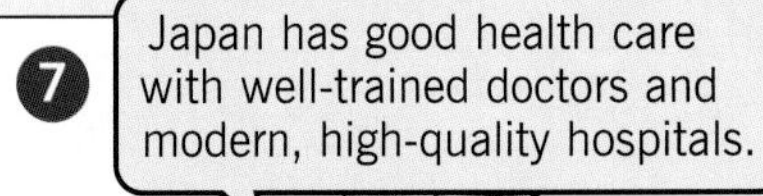

8

9

10 *Wealth*

GNP per capita (US$)	
JAPAN	33,819
UK	30,355
KENYA	444
CHINA	1100
USA	36,924
INDIA	555
ITALY	25,527

11 *Education*

% adults able to read and write	
JAPAN	99
UK	99
KENYA	69
CHINA	70
USA	98
INDIA	50
ITALY	97

12 *Food*

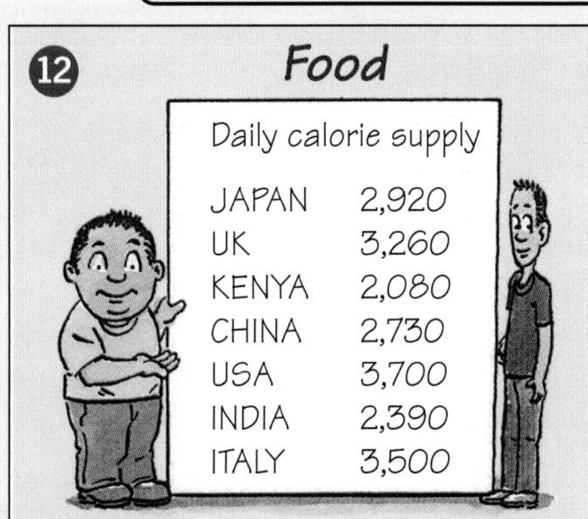

Daily calorie supply	
JAPAN	2,920
UK	3,260
KENYA	2,080
CHINA	2,730
USA	3,700
INDIA	2,390
ITALY	3,500

What is the development problem?

What is this unit about?

This unit is concerned with development, a process of growth and change which can help countries progress and become better places in which to live. The unit is in two parts. The first part looks at the causes of differences in development and the second considers ways of achieving development by the use of aid.

In this unit you will learn about:

- **the effects of population growth, jobs and trade on development**
- **how development is not evenly spread**
- **the benefits and problems of giving aid.**

Why is world development an important topic?

Learning about development can help you understand the problems facing millions of people around the world and appreciate the need for rich countries to help the poor countries to progress and improve their standard of living. It can also help you become a **global citizen**, interested in the state of the world, aware of the problems and willing to do your bit to solve them.

This unit will help you to:

- understand a major world problem
- be aware of the problems that a lack of development causes
- appreciate the need to help people who are less well-off than yourself
- know what you can do to help.

A Niger

B China

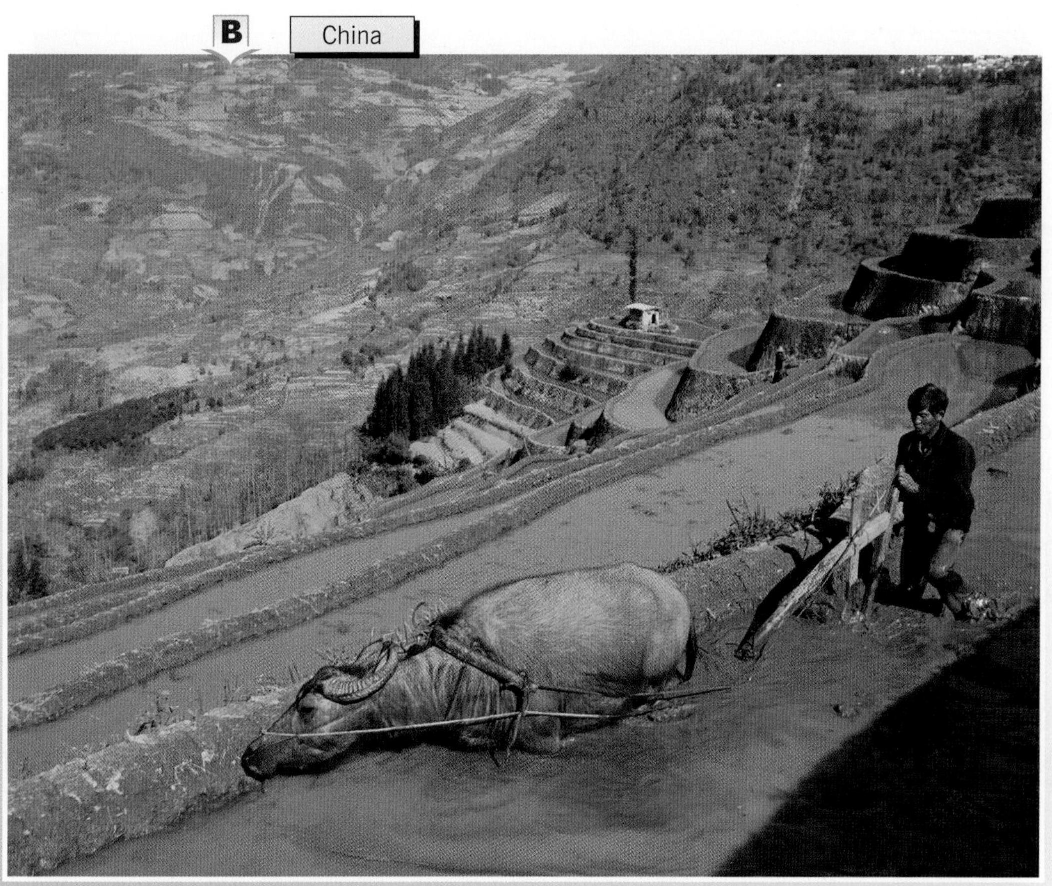

- For each photo, suggest what problems the people face. How might these problems be reduced?
- Which photo do you think best shows a less developed country? Give reasons for your answer.

C India

D Kenya

Where in the world ...?

Activities

1. The world's largest cities and longest rivers are marked on the map. See if you can find them.

2. Complete the crossword by solving the following clues. All of the answers can be found on these two pages.

Across
1 South America's highest mountain
2 A mountain in Africa
3 South American city
4 River in Africa
5 World's highest mountain
6 Ocean east of Africa
7 Aconcagua is in these mountains
8 Everest is in these mountains
9 City on the River Yangtze

Down
1 South American river
10 Mountain range
11 Ocean east of America
12 Sometimes called Antarctic Ocean
13 River in North America
14 Important city in USA
15 Russia's capital city
16 City on River Nile

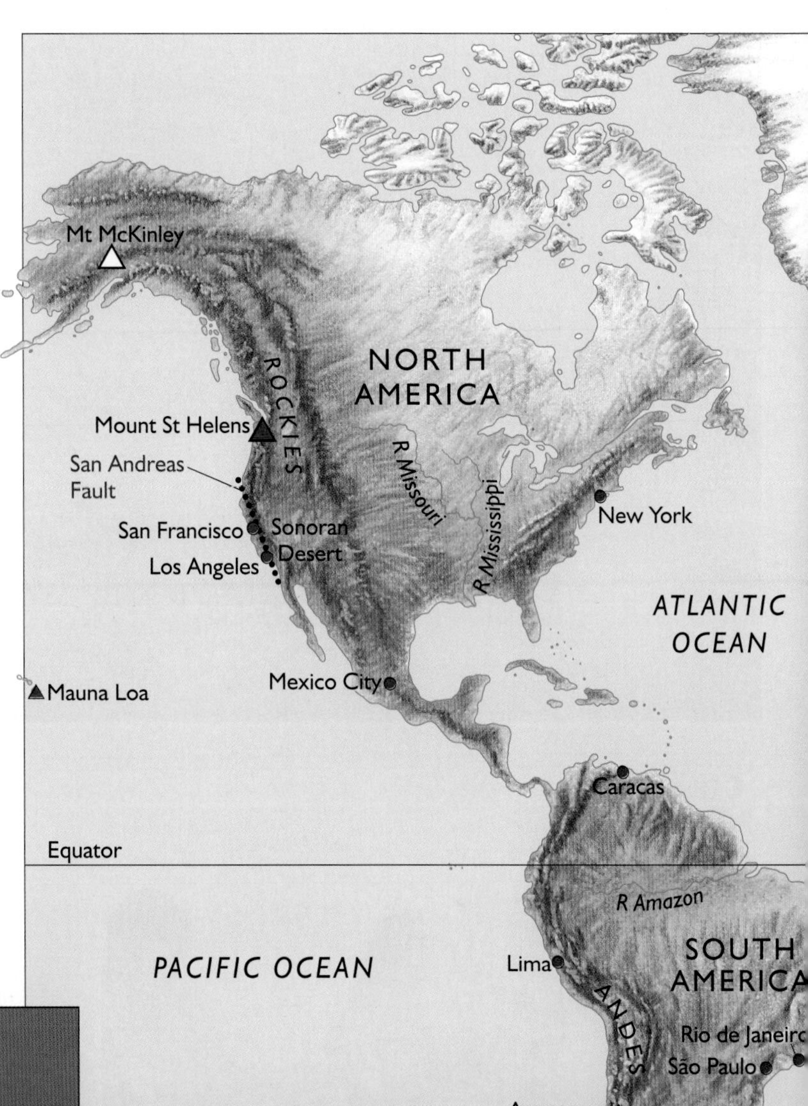

World's largest cities (estimates 2005)

City	*Country*	*Population in millions*
Mexico City	Mexico	25.6
São Paulo	Brazil	22.1
Tokyo	Japan	19.0
Shanghai	China	17.0
New York	USA	16.8
Kolkata	India	15.7
Mumbai	India	15.4
Beijing	China	14.0
Los Angeles	USA	13.9
Jakarta	Indonesia	13.7

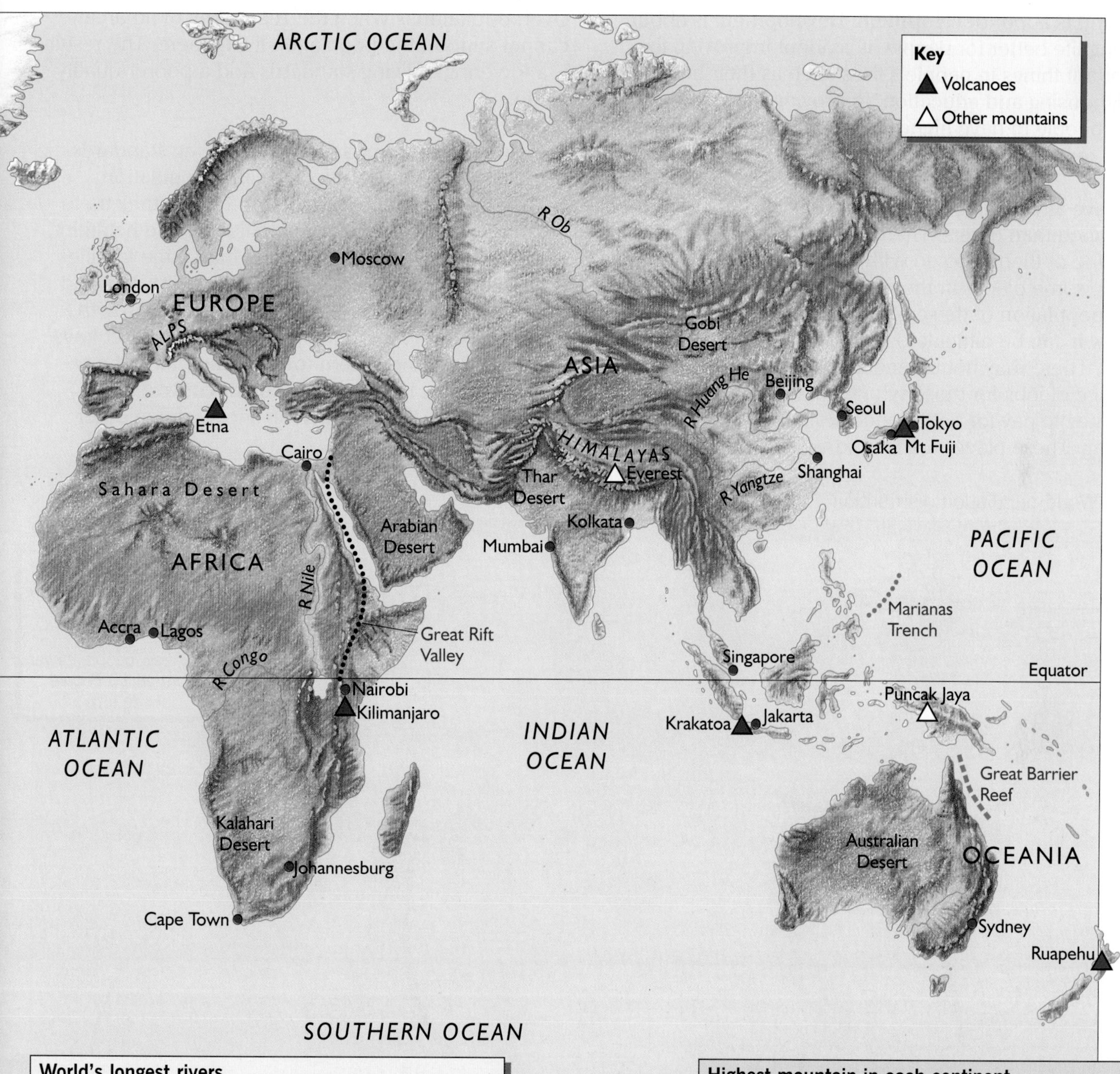

World's longest rivers

River	*Continent*	*Length km (miles)*
Nile	Africa	6,680 (4,150)
Amazon	South America	6,280 (3,900)
Mississippi-Missouri	North America	6,120 (3,800)
Yangtze	Asia	5,470 (3,400)
Ob	Asia	5,150 (3,200)
Huang He	Asia	4,670 (2,900)
Congo	Africa	4,670 (2,900)

Highest mountain in each continent

Continent	*Mountain*	*Height metres (feet)*
Asia	Everest	8,848 (29,028)
Africa	Kilimanjaro	5,895 (19,340)
South America	Aconcagua	6,960 (22,634)
North America	McKinley	6,194 (20,320)
Antarctica	Vinson Massif	5,140 (16,860)
Oceania	Puncak Jaya	4,884 (16,502)
Europe (not Russia)	Mt Blanc	4,807 (15,770)

Too many people?

This unit is about **development**. Development is about making life better for people. It is about improving the important things in people's lives such as their health, food, housing and education. Improving these things helps people to have higher **living standards** and a better **quality of life**.

You have already learnt that some places are less developed than others. There are many reasons for this. One of them is to do with population. As map **A** shows, some places in the world are very crowded. If the population of these places is also growing very quickly it can be difficult to provide for everyone's needs. There may not be enough food to go round, a shortage of jobs for those who want them, and a lack of money to pay for better education, health care or housing. These places are said to be **overpopulated**.

Overpopulation is when the **resources** of an area cannot support the population living there. The result is a lowering of living standards and a poorer quality of life.

To help development and improve living standards, some countries make efforts to limit population growth by reducing birth rates. To do this they try to persuade people that small families can be healthier, better fed, wealthier and happier than large families. China, the world's most populated country, has had the most success in controlling population growth. Here a mixture of education, benefits and strict laws in the 1980s helped them towards their aim of **one-child families**. Unfortunately, **birth control** programmes have not been so successful in other countries.

A World population distribution

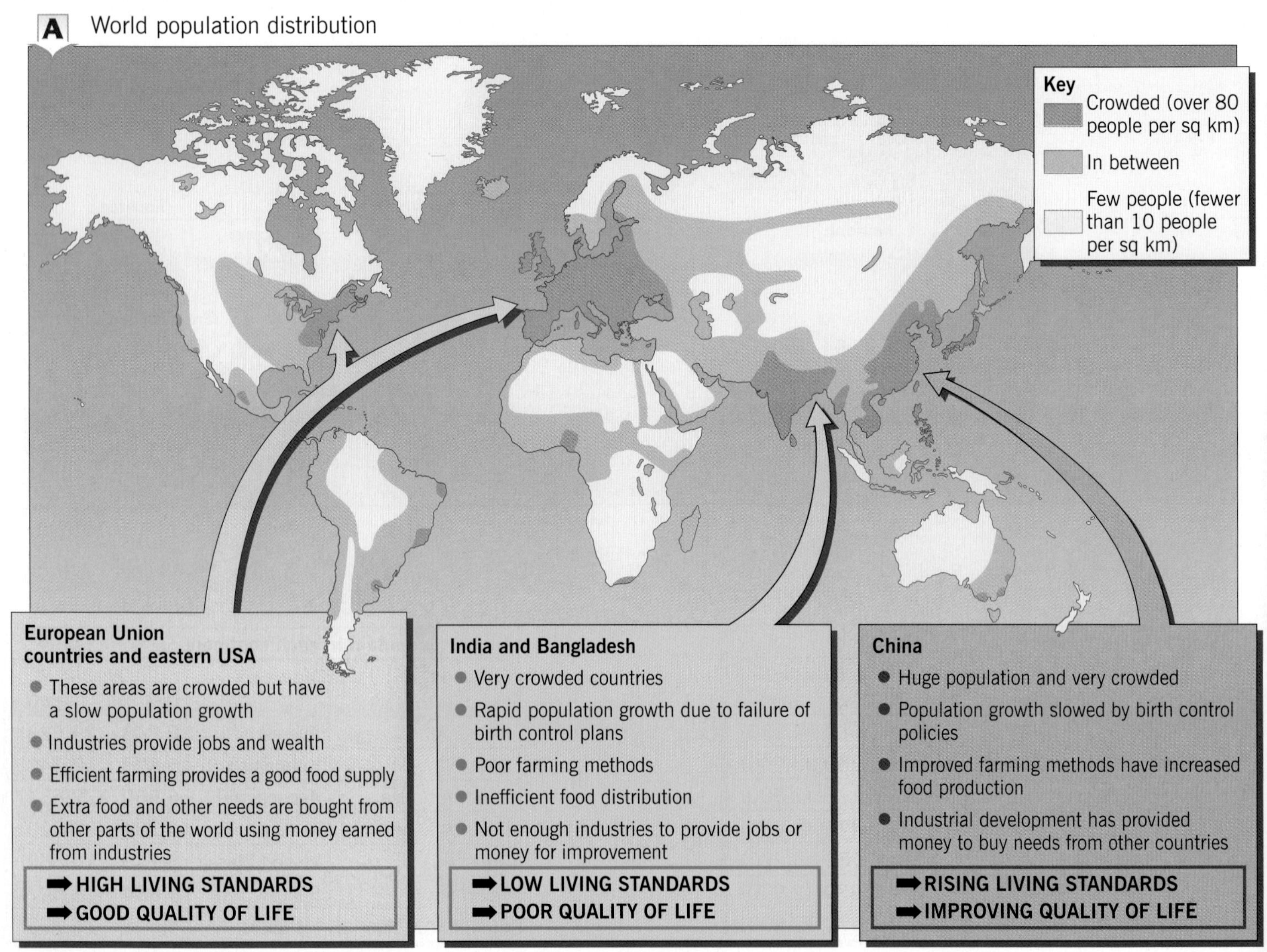

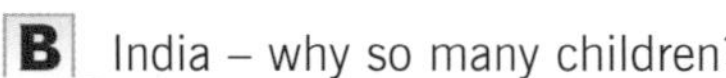

B India – why so many children?

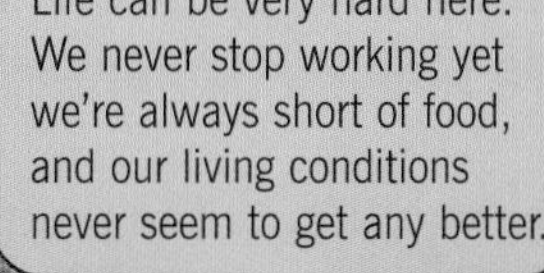

C China – why the one-child family?

Activities

1 Look at cartoon **D**.
Imagine that the number of people in your classroom was suddenly doubled or even trebled.

a What problems would it cause?
b What shortages would there be?
c How would it affect your learning?

2 **a** Write out the meaning of **overpopulation**.
b Give three problems that may result from overpopulation in a country like India.

D

3 Look at sketches **B** and **C** which give some reasons for population size in India and China. Copy and complete table **E** by putting a tick or a cross in each column.

E

	India	China
Children needed to do work		
Children needed for old age		
Pension scheme for old age		
Good health care		
Disease problems		
Large families		
Improved conditions		
Only slow improvement		

4 What are the advantages of having small families?

EXTRA

Explain how crowded places like the European Union and eastern USA have high living standards and a good quality of life.

Summary

Some countries are crowded and have a rapid population growth. This can cause problems for people and slow down development.

How do jobs affect development?

The jobs people do can be divided into three different types. These are **primary**, **secondary** and **tertiary**. The number of people working in these different types of job varies from place to place and changes with time.

The proportion of the population working in primary, secondary or tertiary jobs in any place is called the **employment structure**. The diagram below shows how the employment structure changes as a country develops.

Map **B** shows some differences in employment structures. These are constantly changing, however. For example, 200 years ago the UK had an employment structure similar to that of present-day India.

Since then changes in farming methods and the growth of manufacturing industries and services have altered that structure. The UK is now one of the world's more developed countries.

B World employment structures

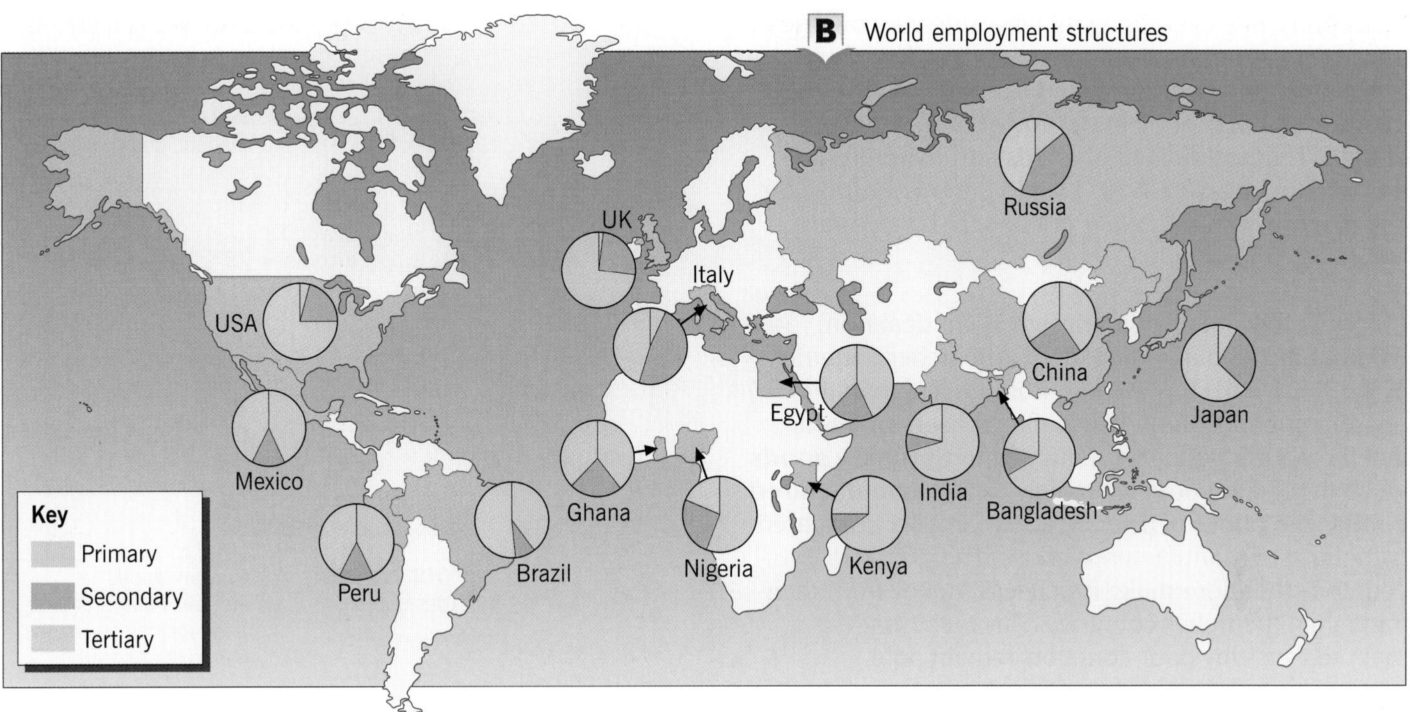

Activities

1 Give the meaning of the term **employment structure**.

2 Four of the following six statements are correct. Write out the correct ones.

- Employment structures are the same everywhere.
- Employment structures vary from place to place.
- Employment structures change from time to time.
- Poor countries have few workers in primary industries.
- Poor countries have many workers in primary industries.
- Rich countries have few workers in primary industries.

3 Using map **B**:

a Name the four most developed countries (those with very few primary jobs).

b Name the eight least developed countries (those with a lot of primary jobs).

4 Table **C** shows job types in Britain for 1790, 1890 and 2005.

a Draw three pie charts to show the information. Use the same colours as those on map **B** and add a key.

b Label each of the pie charts with one of the following:

- A more developed stage
- Beginning to develop
- Early stages of development.

c Describe the changes shown by the pie charts.

C

Year	Primary (%)	Secondary (%)	Tertiary (%)
1790	75	15	10
1890	4	56	40
2005	2	25	73

EXTRA

Describe and give reasons for the living standards that are likely in:

- Bangladesh and Kenya
- Italy and Japan.

Summary

There is a link between the employment structure of a country and its level of development. Poor countries have a larger primary workforce than rich countries.

How does trade affect development?

No country has everything that it wants. All countries have to buy from and sell to each other. They **buy** things that they need or would like to have. They **sell** things to make money to pay for what they have bought. The exchanging of goods and materials like this is called **trade**. Goods sold to other countries are called **exports**. Goods that are bought by a country are called **imports**.

Unfortunately not all countries get a fair deal from world trade. As diagram **A** shows, the richer countries of the North earn much more from trading than the poorer countries of the South. One reason for this is that the poorer countries mainly export **primary goods** while the richer countries mainly export **manufactured goods**. The prices of primary goods are low compared with those of manufactured goods. The poorer countries therefore make much less money from their trade than the richer countries. This unfair trade is a main reason why poor countries remain poor.

A North ↔ South exports

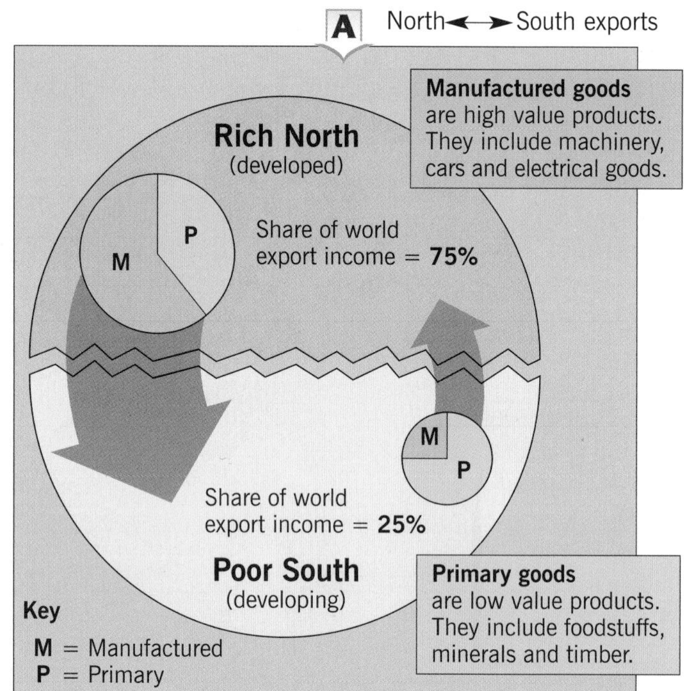

B Loading timber from the Amazon Forest, Brazil

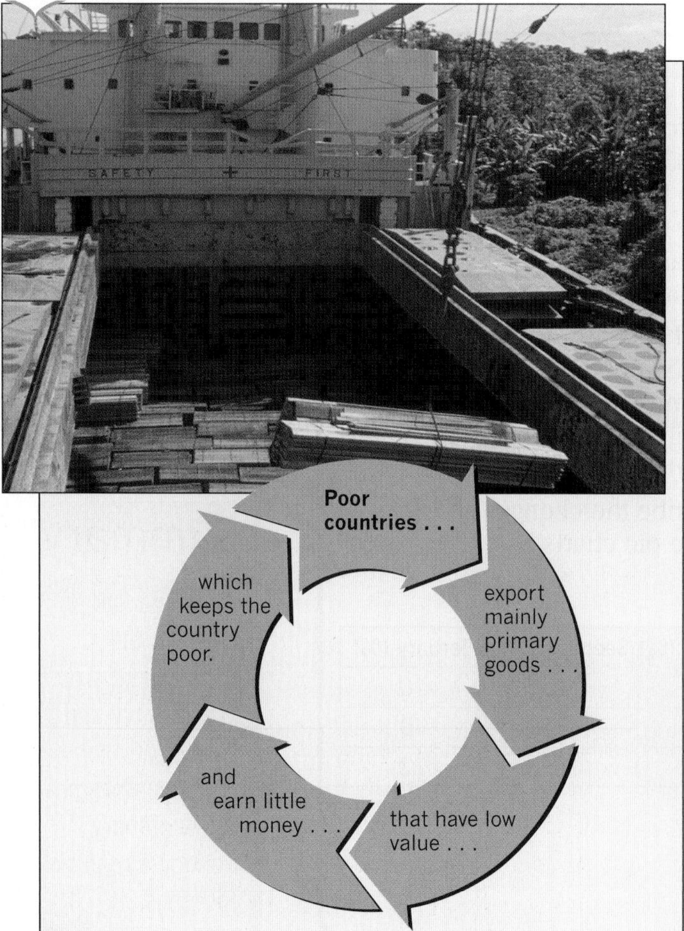

C Cars being loaded onto a ship in Japan

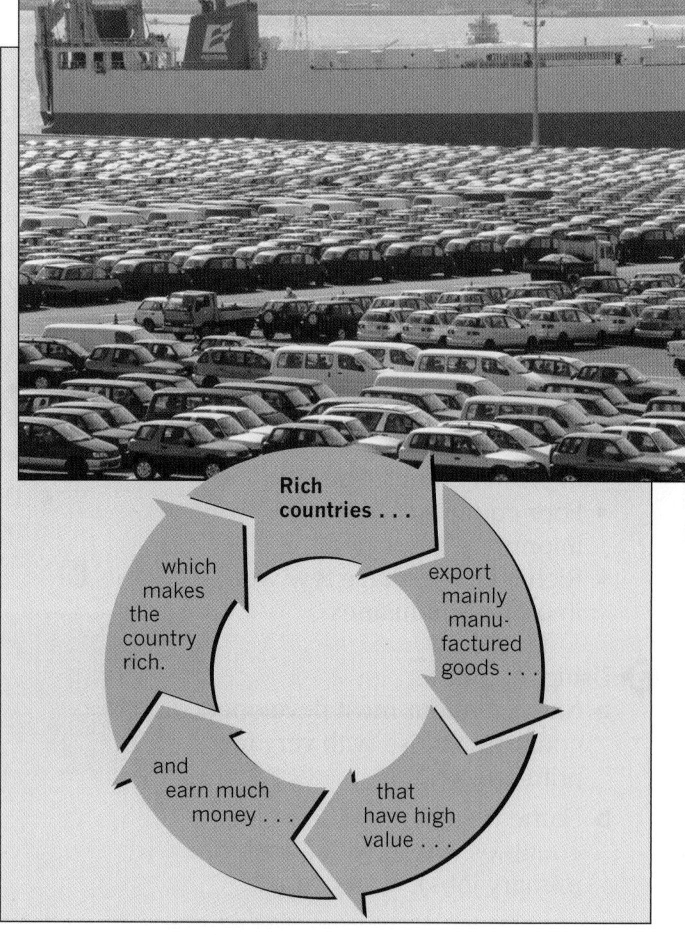

Diagram **D** shows another problem faced by developing countries. Many of them rely heavily on just one major export. Ghana, for example, earns 80 per cent of its income from selling cocoa. If there is a good crop and the price for cocoa is high then Ghana will prosper. If, however, the crop fails or the price in the world market falls, the country may struggle to survive because it is so dependent on that one export.

There have been many attempts to keep the prices of primary goods high to try to protect countries like Ghana. Oil-producing countries, for example, have joined together and successfully controlled oil prices. Nigeria and Egypt are countries that have benefited from this success. Unfortunately most other attempts to control prices have failed.

D Single export countries

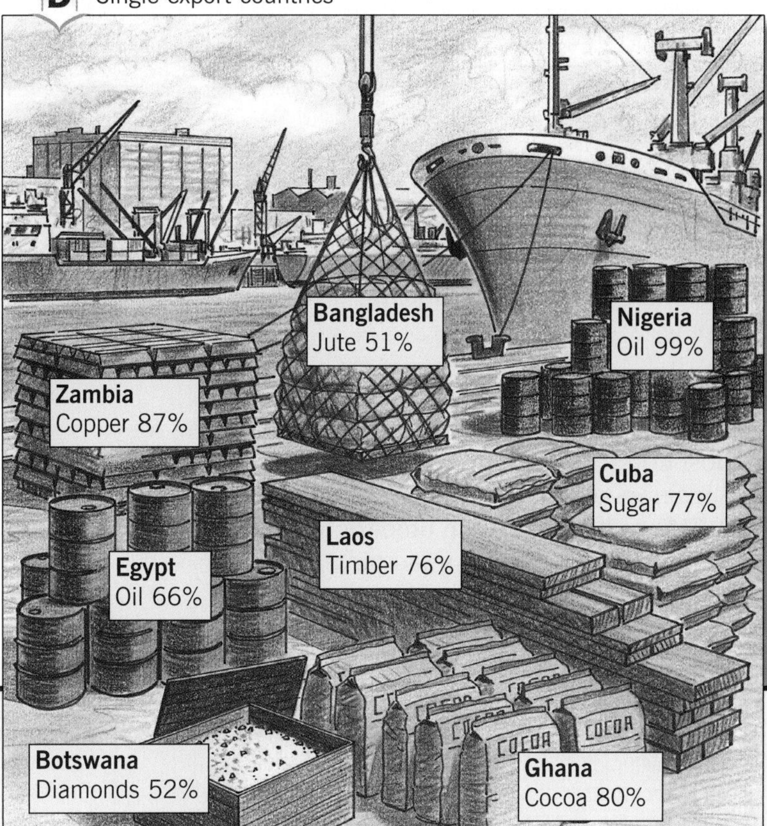

Activities

1 **a** Make a copy of table **E**.

Country	GNP	Primary exports	Manufactured exports

b List the eight countries from **F** in order of their wealth. (Highest GNP first)

c List the exports of each country in either the **Primary** or **Manufactured** columns.

d Shade lightly in yellow the less developed countries. (GNP less than US$ 3,000)

e Write a sentence to describe what your table shows about the exports of:
- the more developed countries
- the less developed countries.

F

Country	GNP (US $/person)	Main exports
Brazil	2,700	Coffee, machinery, meat
Egypt	1,062	Oil, fruit, vegetables
Ghana	354	Cocoa, aluminium, timber
Italy	25,527	Machinery, clothing
Japan	33,819	Electrical goods, cars
Kenya	444	Coffee, tea
UK	30,355	Machinery, chemicals
USA	36,924	Machinery, chemicals

2 Make a larger copy of diagram **G** and add the following information to the correct boxes.
- Crop failure
- Loss of income
- Decline in living standards
- Reliance on one main export
- Fall in price

G

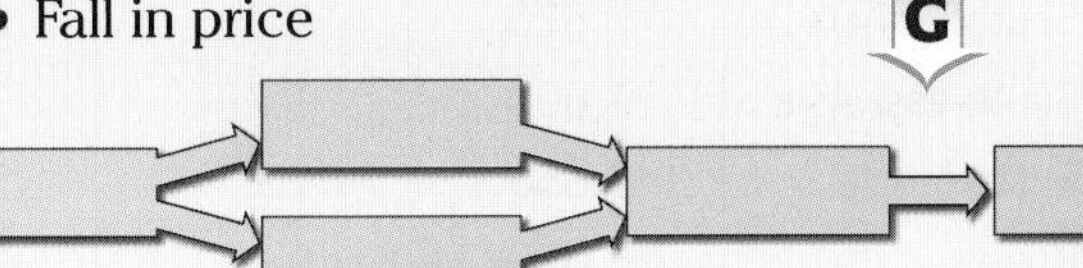

3 **a** From diagram **D** name the countries that depend on one product for:
- over three-quarters of their exports
- between half and three-quarters of their exports.

b Why do countries that depend mainly on export often have money problems?

Summary

Trade is important in the world because it helps countries share resources and earn money. Rich countries gain more from trading than poor countries.

Is development spread evenly?

Development is about progress and making life better for people. As we have seen, however, all countries are different. Some are rich, have high standards of living and are said to be developed. Others are poor, have low standards of living and are described as developing.

There are many different ways of measuring development. Some of these are shown in drawing **A**. The most common and easiest way is to measure wealth using a country's **Gross National Product (GNP)**. This, when used with other economic and social measures, can give a good picture of a country's level of development.

But where in the world are the developed countries and developing countries located? Is there a pattern to their distribution or are they simply scattered randomly around the world?

The best way to find out is by using a map and plotting on it each country's level of development. This can be done but would be awfully complicated – there are over 200 countries in the world! Table **B** below has simplified the task by reducing the world to just 16 different regions. Remember, though, that the statistics provided are just averages and only give an approximate guide to each region's level of development.

A

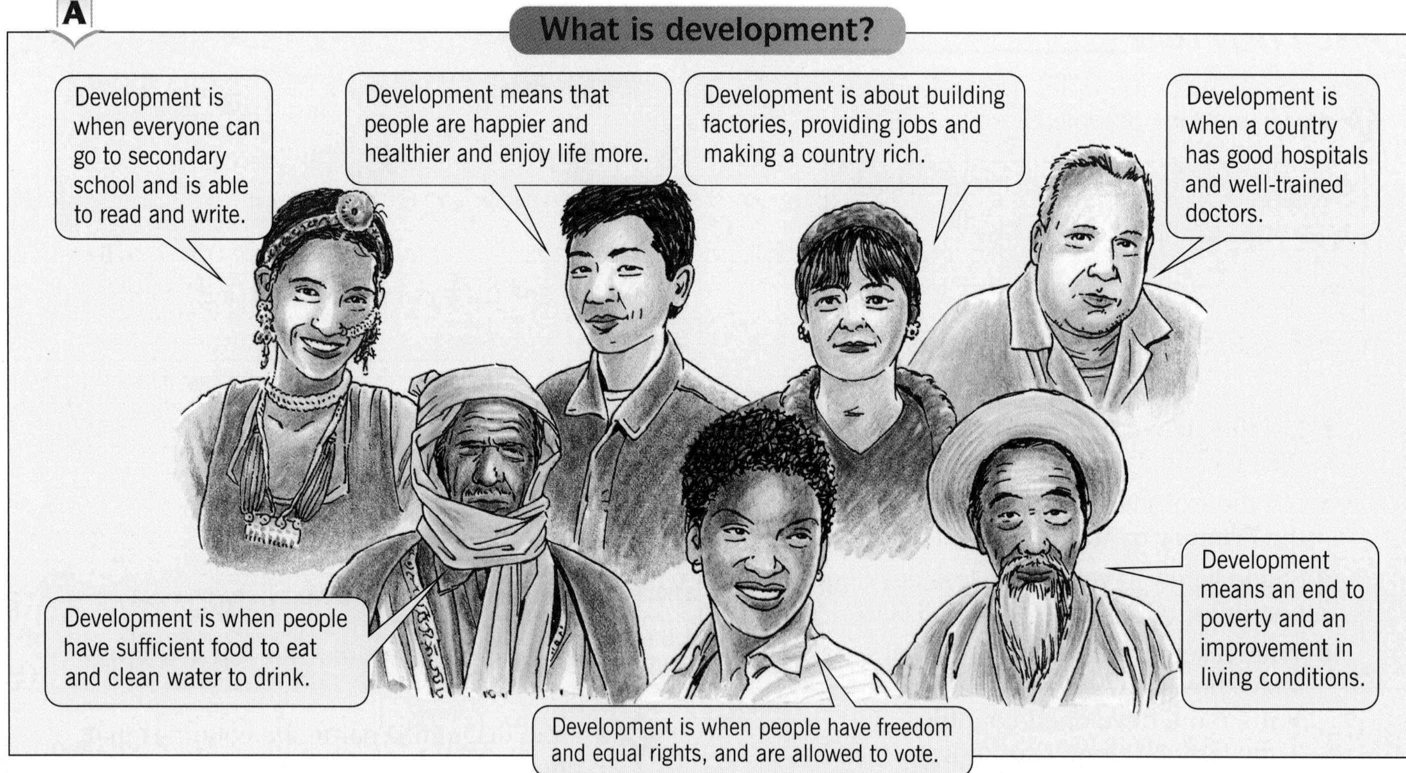

B

World region	GNP per capita (US$)	Population increase (%)	Primary jobs (%)
North America	24,200	0.8	3
Central America	3,100	2.5	10
South America	2,950	1.7	32
Northern Africa	1,050	2.6	18
Western Africa	420	3.1	52
Eastern Africa	320	3.0	62
Southern Africa	2,640	1.8	21
Australia/New Zealand	23,750	0.8	6

World region	GNP per capita (US$)	Population increase (%)	Primary jobs (%)
Western Europe	23,400	0.2	5
Eastern Europe	2,160	0.1	16
Western Asia	4,100	2.8	21
Southern Asia	520	2.3	31
South-east Asia	1,600	1.7	71
Eastern Asia	1,850	1.1	73
Russia	3,050	0.6	20
Japan	33,819	0.3	2

C World regions base map

Activities

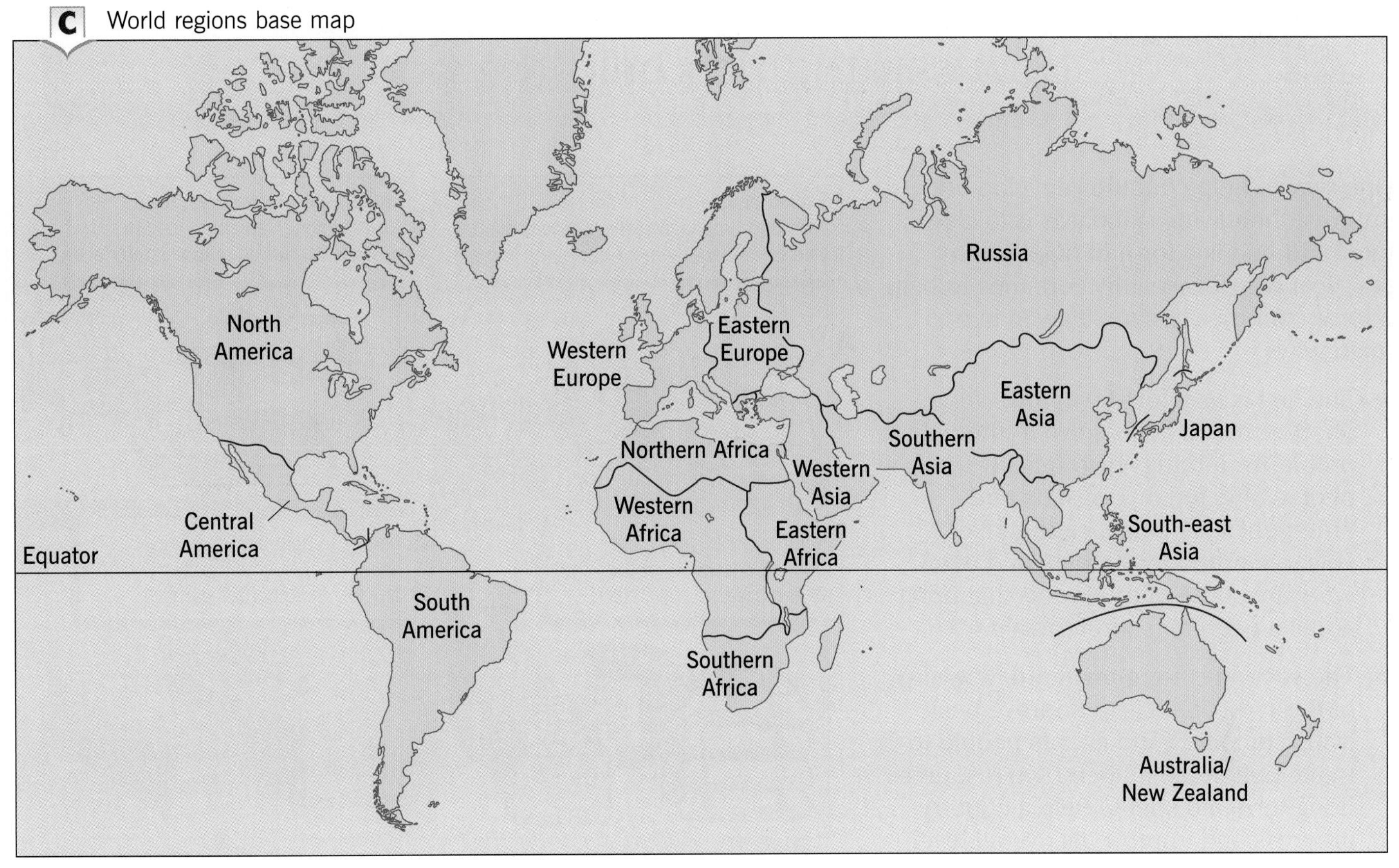

1 Give the meaning of the following terms. If you are not sure of any, check the Glossary.

- **a** Gross National Product per capita
- **b** Economic indicators of development
- **c** Social indicators of development

2 Do you think that you live in a highly developed country? Give reasons for your answer using both economic and social examples.

3 You will need three copies of map **C** for this activity.

- **a** Colour in each map using the information in table **B**. The first map will show GNP, the second one population increase, and the third one primary jobs. Use the keys shown in drawing **D** to work out your colours. Add a title and key to each map.
- **b** Look carefully at each map and describe the pattern shown. Are the developing areas scattered all around the world or grouped together? Are some continents more developed than others? Is the north richer or poorer than the south?

D

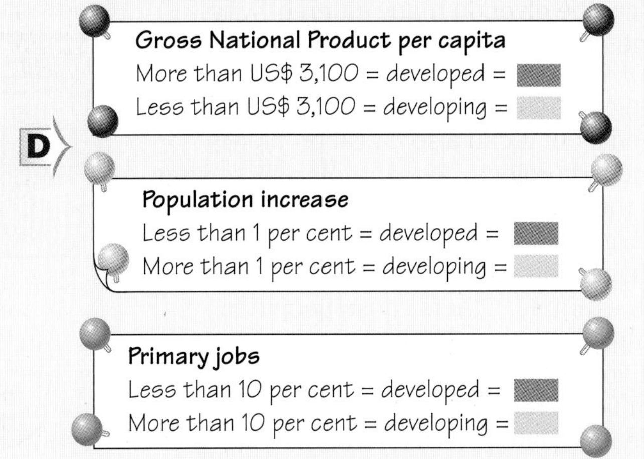

- **c** Now look at your three maps together. Can you see a line that divides the rich and poor regions? If you can, draw the line on your maps. Describe the way the line has divided up the world.

Summary

Development is not spread evenly around the world. Development indicators show that the north is generally richer and more developed than the south.

How can the rich help the poor?

One way to help countries develop and improve their living standards is to give them **aid**. Aid is a form of help. It is a practical way for wealthy countries to help poorer countries. It can be given in two main ways.

- The first is as **short-term aid**. Short-term aid helps solve immediate problems. It brings help quickly to people affected by disasters and emergencies. Floods, earthquakes, volcanic eruptions, famine and even wars are some of the events that bring about a need for short-term aid.
- The second is **long-term aid**. The aim of this type of aid is to improve basic living standards and enable people to make better use of their own resources. Long-term aid should help a country progress and improve its overall level of development.

Aid can be given in many different ways. Some of these are shown in diagram **A**.

Giving help to others can bring many benefits but it can also cause problems. Some aid projects, for example, are so large that they damage the environment and are too big and complicated for local people to manage. The Aswan Dam in Egypt is one such example. Others cause people to change their lives too much and spoil the traditions of the area. Some forms of aid even fail to reach the people for whom they were intended. In Somalia, Sudan and Niger, for example, some of the food aid sent there never reached the millions of people who were dying of starvation. This was partly due to a lack of transport but also in some areas because of civil war.

Great care has to be taken in providing countries with the right kind of aid. A well thought out and carefully planned programme can help to provide the building blocks for a country's future. Cartoon **B** shows the kind of aid which is most likely to bring benefits to a country and help its poorest people.

Activities

1 **a** Draw a table and sort the newspaper headlines **C** into two columns headed **Short-term aid** and **Long-term aid**.

b Choose two headlines from each column and suggest what aid might be most helpful to that country. Diagram **A** will help you.

2 **a** Down the left hand side of your page write a list of six rules that you think all aid should keep to. Put them in the order you think most important. Diagram **B** will help you.

b Look at **D** below showing a water supply scheme in Kenya. For each rule, answer *Yes*, *No* or *Partly* to describe whether the scheme is keeping to that rule.

c Give the scheme a mark out of ten and write out a comment on its success.

C

Earthquake wrecks Mexico City

Record debts hit Brazilian banks

China seeks help for farming problems

Crops lost in Bangladesh flood

Water supply blamed for Ghana disaster

Massive new health scheme planned for India

D

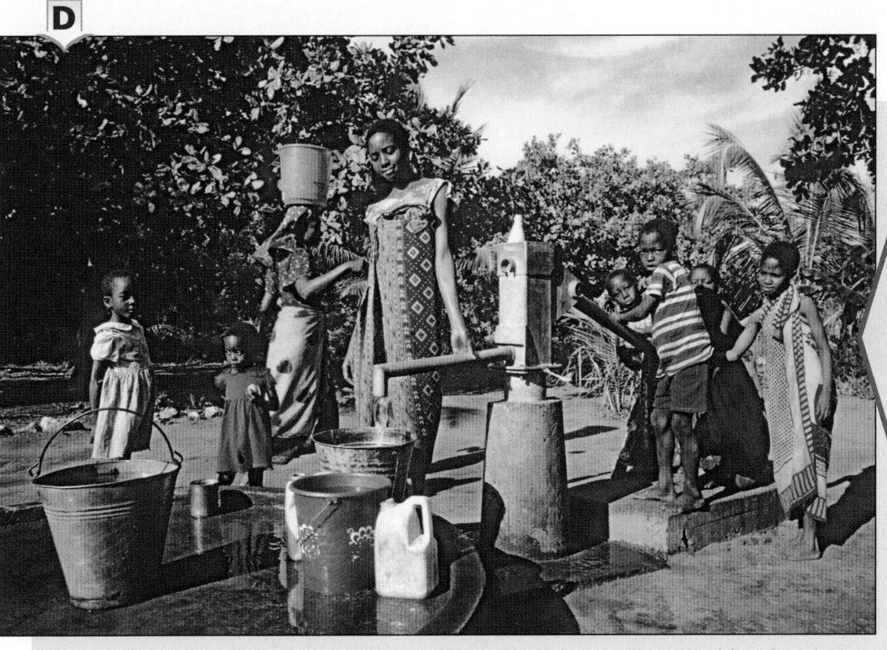

- Simple **technology**; easy and cheap to run and repair
- Provides direct help for the people most in need
- Advisers help local people to make best use of the new water supply
- Water supply can become a village meeting point
- Small-scale project with no bad effects on the environment
- Does not change local way of life

An example of **sustainable development**: a simple borehole well with a hand pump, which requires no servicing, supplies clean water to a small village in Kenya.

EXTRA

Imagine that you are to help organise the provision of a safe water supply to a country in Africa. Which eight of the following would be most useful to you? For each one that you choose, say how it would be used.

• *money* • *farmer* • *drilling equipment* • *nurse* • *lorry* • *laboratory* • *water pipes* • *mechanic* • *tractor* • *chemist* • *water pump* • *geologist* • *teacher*

Summary

Aid is a form of help usually given by the wealthy areas of the world to the poorer areas. Aid schemes should be sustainable. If they are planned carefully they can help provide a better life for people living in poorer areas.

Should we give aid?

Aid comes from many different sources. In 2005 the British government gave around £3,200 million of aid to developing countries. **Charities** like Oxfam and Save the Children spent a total of over £120 million mainly funding small schemes and giving emergency aid. Large international organisations like the United Nations, the World Bank and the European Union provided more than £900 million of help to the poorer countries and regions of the world.

There are arguments as to whether we should give aid. Some people think that aid can be damaging and that people should help themselves. Others point out that we all live in the same world and we all rely on each other for our survival. We must therefore help each other and try, as far as possible, to improve the quality of life for everyone. What do *you* think? Look carefully at diagram **A** and try to decide whether we should or should not give aid.

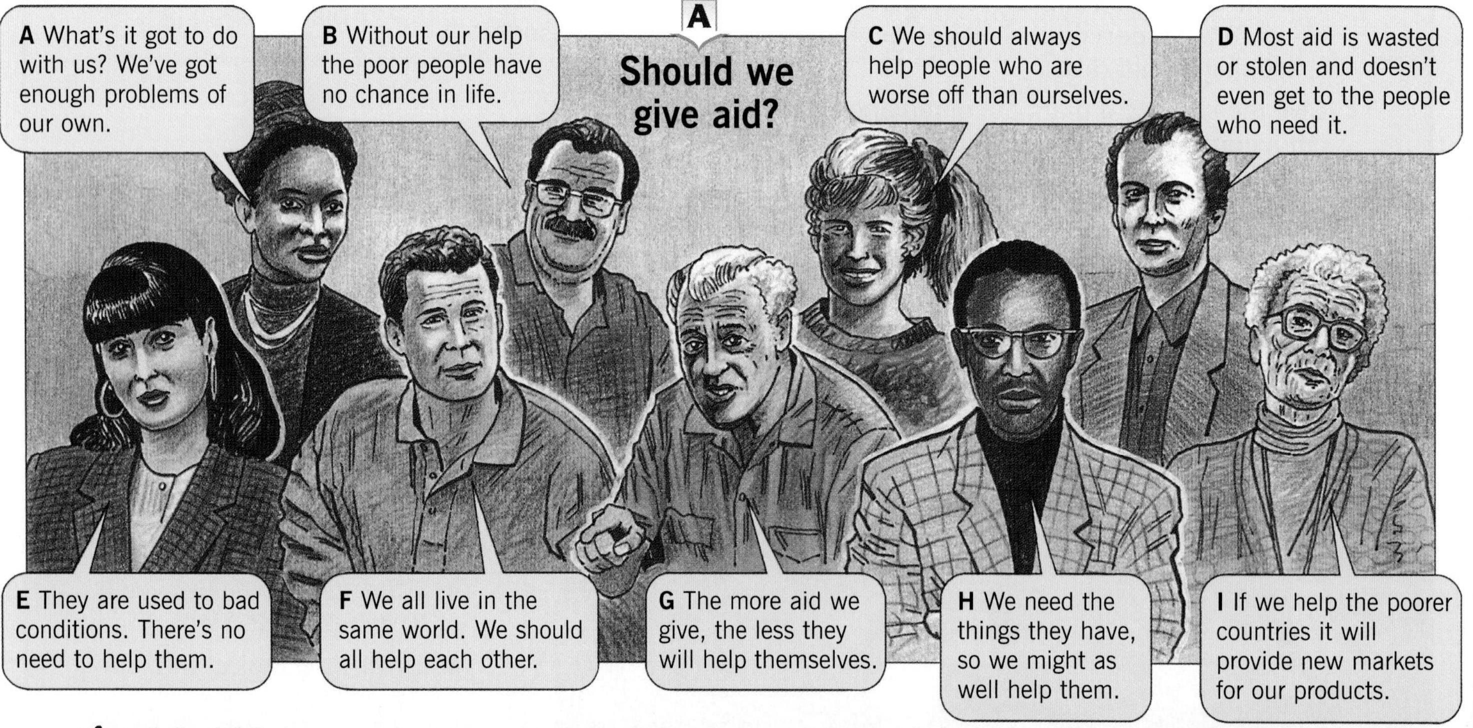

Activities

1 Look at diagram **A**.

a Give the letters of the speech bubbles that are:
- **for** giving aid
- **against** giving aid.

b Write out the two speech bubbles that you think are the best argument:
- **for** giving aid
- **against** giving aid.

c Do you think we should give more aid, less aid or no aid to people who live in countries that have a very low standard of living? Give reasons for your answer.

2 a With a partner, play the game on the opposite page.
- Use a dice or spinner for each move.
- Follow the instructions for moves backwards or forwards.

The winner is the first to reach the end of the road with the exact number.

b Play the game again. This time write down the **problems** and the **help** that you meet on the way. Do this in two columns. Underline the events which were affected by aid.

Summary

There are great differences in levels of development around the world. It has become increasingly difficult to improve living standards for people in the poorer countries.

World development enquiry

ActionAid is a charitable organisation which was founded in 1972. It is the UK's fourth largest overseas development agency, and works in partnership with poor people in more than 30 countries across Asia, Africa and South America. Its aims are to help these people improve their standard of living and gain access to their basic rights.

ActionAid has worked in Kenya since 1976. Projects are mostly in rural areas, but it also runs urban projects in squatter settlements – many people from the countryside migrate here in search of a better life. ActionAid's approach is to support local community groups in setting up and running their own projects.

In this enquiry you should imagine that you work on the Project Team in ActionAid Kenya's Nairobi office. Your job is to decide how best to use the funds raised in the UK and elsewhere for ActionAid's work in Kenya. You have £50,000 to spend and have been asked to look at three proposals put forward by local community groups.

A

actionaid

- Encourages schemes set up and run by local groups
- Supports appropriate and sustainable long-term solutions to local problems in less developed countries
- Supports projects that enable people to understand and gain access to their basic rights

B

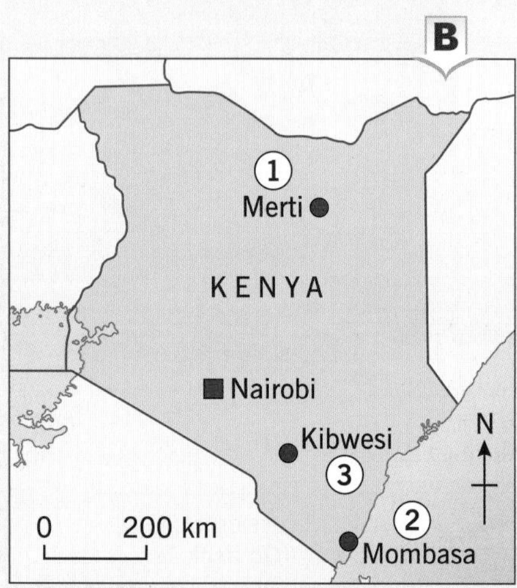

C

Factors to consider	Scheme 1	Scheme 2	Scheme 3
Provides nearby clean water supplies			
Increases food production			
Reduces disease and infection			
Improves health care			
Improves ability to earn money			
Helps people learn to read and write			

What is the best way to help poor people improve their lives?

1 **a** Copy table **C** which shows some factors that have to be considered when choosing which schemes to support.

b Read the proposals in **E** carefully. Show the advantages of each scheme by putting a tick in columns 1, 2 or 3 in table **C**. More than one column may be ticked for each factor.

c Add up the ticks to find which scheme has the most advantages.

d Decide which two schemes you would choose. The two with the most advantages will be the best. Consider also how each scheme matches ActionAid's main aims shown in drawing **A**.

2 Working with the local group, decide how to spend the £50,000. To do this, copy table **D** and list the items and costs for your two chosen schemes. Decide how many of each item you want and work out the total costs.

D

Item	Cost	Number	Total cost

3 Write a report for the Kenya Country Director giving details of your decisions. Describe the two schemes and explain how they will improve the quality of life for local people. Link your explanations to ActionAid's aims.

E Proposed development schemes

Scheme 1

Merti district is home to 18,000 people in the dry, isolated north of Kenya. The main health centre is empty and falling down, and most people have no health care.

'We want to re-open the centre and set up mobile clinics to provide basic medical needs, vaccinations and advice on nutrition, hygiene and childcare. If we can have help with materials, equipment and medical supplies, local volunteers will rebuild the health centre and convert some old jeeps for use as mobile clinics. With training, our community groups can take over health care and collect low-cost fees so that medicines can be re-stocked.'

Rebuild and re-stock health centre	**£10,000**
Set up mobile clinic	**£2,000 each**
Provide health worker training	**£500 per person**

Scheme 2

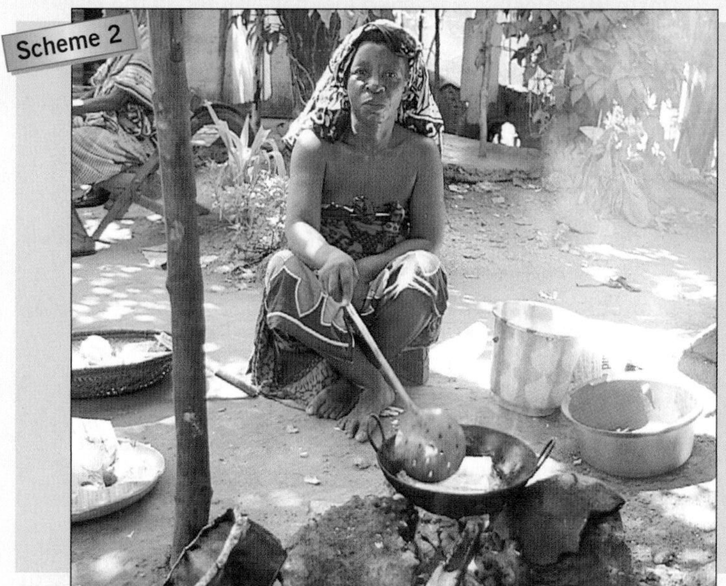

Ziwa La Ng'ombe is a shanty settlement of about 70,000 people on the outskirts of Mombasa.

'Many of us are women on our own with young families to support. Most of us are illiterate, have few job skills and no money. We would like to set up small savings groups of 15 to 20 women each, so that we can put money aside for expenses like school fees and take out loans to set up small businesses like dressmaking. If ActionAid could also set up adult literacy and skills training classes, this would build our confidence and increase our earning ability. We could buy better food, afford to see the doctor and help our children through school.'

Savings group (training and start-up fund)	**£500 each**
Skills training centre	**£1,000 per trade**
Literacy unit	**£1,500**

Scheme 3

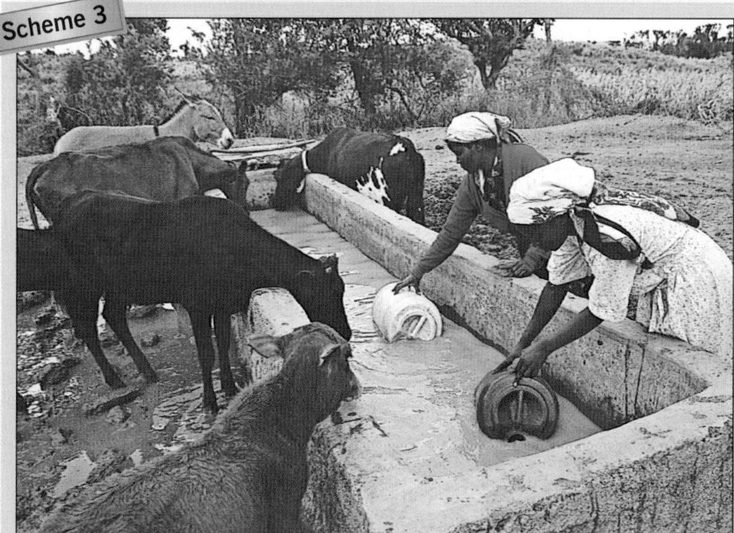

The **Kibwesi district** is home to 90,000 people. The area suffers severe water shortages and is short of food. Many people suffer illness caused by dirty drinking water.

'We want to improve our water provision by digging several wells and building simple water storage systems. We would set up vegetable plots near the wells, build fenced cattle troughs for our livestock and dig ponds for fish farming. These ideas would improve our diet, and extra produce could be sold at the market to make money. Local groups would organise work parties if ActionAid could supply the necessary materials and training.'

Water provision	**£2,000 per unit**
Crop farming (materials and training)	**£1,000 per plot**
Fish farming (materials and training)	**£2,000 per pond**

Glossary and Index

Acid rain Rainwater containing chemicals that result from the burning of fossil fuels. *10*

Active volcano A volcano that has erupted recently and is expected to erupt again. *32–35, 44–45, 90, 110, 112–113*

Adapt Plants that have learnt to live with the temperature, rainfall and soils of an area. *14–15, 18*

Aid Help usually given by rich countries, international agencies and voluntary organisations to poorer countries. It may be given short-term or long-term. *81, 121, 136–141*

Altitude The height of a place above sea-level. *7, 13, 16, 91*

Ash and dust Fine material thrown out by a volcano when it erupts. *32–34, 95*

Balance of trade The difference in value between imports and exports. *121*

Biome A very large ecosystem such as the rainforest or coniferous forest. *10–11, 25*

Birth control When the birth rate is reduced by artificial methods. Also referred to as family planning. *128–129*

Birth rate The number of live births for every 1,000 of the population per year. *103, 128*

Brand name A name or trademark that is well known and easily recognisable. *70, 72*

Buttress roots Roots that stand above the ground to support large trees. *14*

Canopy An almost unbroken top layer of trees that acts like a roof over the tropical rainforest. *14–15*

Charities Voluntary organisations that provide help for those in need. *138–141*

Climate The average weather conditions of a place taken over many years. *6–10, 12–14, 16–18, 56, 91, 99, 112–113*

Commercial farming The growing of crops and rearing of animals for sale and profit. *100*

Conservationists People who care for and look after resources and the environment. *54*

Convectional rainfall Rain that is produced when air rises after being warmed by the ground. *8, 13*

Core The central part of the earth. *30*

Crater A roughly circular opening at the summit of a volcano. *32–33, 35, 44*

Crust The thin outer layer of the earth. *30*

Deforestation The cutting down or burning of trees to clear large areas of land. *20, 22–23*

Delta A flat area of fertile land deposited by a river where it enters the sea. *98*

Developed country A country that has a lot of money, many services and a high standard of living. *104, 122, 130, 132, 134*

Developing country A country that is often quite poor, has few services and a low standard of living. *49, 75, 77, 104, 122, 130, 132–134*

Development Development involves changes that usually bring improvement and growth. Countries can be at different stages of development depending on how 'rich' or 'poor' they are. *104–105, 122–125, 128–141*

Dormant volcano A volcano that has erupted in the last 2,000 years but not recently. *32*

Drought A shortage of water resulting from an unusually long period of dry weather that may last for months or even years. *16, 18, 91, 99, 102*

Earthquake A sudden movement, or tremor, of the earth's crust. *26–27, 29–31, 36–43, 90, 98, 113*

Economic activities Primary, secondary or tertiary (service) jobs. *100, 102, 130–131*

Economic indicators Measures of development that are based on wealth, e.g. GNP, trade and energy use. *122–123*

Ecosystem A community of plants and animals that live together in the environment. *4–5, 10–11, 24–25*

Emergents Exceptionally tall trees that grow above the canopy level of the tropical rainforest. *14*

Employment structure The proportion of people working in primary, secondary and tertiary activities. *100, 102, 130–131*

Equatorial climate Places near to the Equator that are hot and wet throughout the year. *9, 12–13*

European Union A group of twenty-five European countries working together for the benefit of everyone. *88–89, 106, 138–139*

Evergreen Trees that have green leaves growing throughout the year. *14, 18*

Exploit When a country or group of people take advantage of another country or group of people. *75*

Exports Goods produced in one country that are sold in other countries. *76–77, 116–117, 120–121, 123, 132–133, 139*

Extinct volcano A volcano that is not expected to erupt again. *32*

Fashion The present-day most popular style of clothing. *68–77*

Frontal rain Occurs when warm air is forced to rise over cold air in a depression. *8*

Globalisation The process by which corporations, ideas and lifestyles are spreading around the world with increasing ease. *70–73, 76–77, 79*

Gross National Product (GNP) The wealth of a country: the total amount of money earned by a country in a year divided by its total population. It is a measure of economic development. *71, 122–123, 133–135*

Hardwoods Deciduous trees that grow mainly in the tropical rainforest. *14*

Hazard A natural danger to people and their property and way of life. Hazards include earthquakes, storms, drought and floods. *22, 26–45, 99, 113*

Hydro-electricity Energy obtained from using fast-flowing water. *101, 113, 116*

Imports Goods bought by a country that are produced in other countries. *77, 116–117, 120–121, 123, 132, 139*

Infrared photograph Photo obtained from a satellite that orbits the earth and which records the amount of heat (radiation) given off from different earth surfaces. *95, 112, 118*

Intensive farming Farms that are often small in size but which use either many people or a lot of money (capital). No land is wasted. *95, 100*

Interdependent When countries work together and rely on each other for help. *120–121*

Land use Describes how the land in towns and the countryside is used. It includes housing, industry, farming and recreation. *22, 94, 100–103*

Latitude How far a place is north or south of the Equator. *6, 8, 13, 17, 24*

Lava Molten rock (magma) that usually flows from the crater of an active volcano. *32–35, 44–45, 95*

Lianas Vine-like plants that climb up trunks of trees before hanging downwards from the branches. *14*

Life expectancy The average number of years a person can expect to live. *123*

Literacy rate The proportion of people who can read and write. *105, 123, 134*

Living environment The part of the environment that includes plants, insects and animals. *10*

Magma Molten rock below the earth's surface. *31–33*

Magma chamber Where molten rock is found deep below the earth's surface. *32*

Mantle The layer of the earth beneath the crust and above the core. *30*

Manufactured goods Secondary industry products such as cars, computers and electrical goods. They are usually of high value. *116–117, 121, 132, 139*

Mediterranean climate Places that have hot, dry summers and mild, wet winters. *9, 16–17, 25, 91, 94, 99*

Migrant workers People who move from place to place to find work. *101, 103, 106*

Migration The movement of people from one place to another to live or work. *92, 100, 102–103*

National Park An area of attractive countryside where scenery and wildlife are protected so that they may be enjoyed by both visitors and people who live and work there. *52–55, 64–67, 119*

Natural hazard A great force of nature, such as an earthquake, flood or storm, which is a threat or danger to people and their way of life. *34, 113*

Natural resources Raw materials that are obtained from the environment, e.g. water, coal and soil. *120*

Natural vegetation Vegetation that has not been affected by human activity. *10–11, 14–15, 18, 25*

Non-living environment The part of the environment that includes non–living features such as solar energy, water, air and rocks. *10*

Nuclear power Energy (electricity) obtained from uranium. *116*

One-child family A method of reducing population growth by encouraging parents to limit their family size to one child. *128–129*

Overgrazing Damaging pasture by keeping too many animals on it. Overgrazing can lead to soil erosion. *20, 23*

Overpopulated When there are more people living in an area than that area can support. *128*

Peninsula A piece of land surrounded on three sides by the sea. *94, 97*

Plain A flat, low-lying area of land. *90, 98–101*

Plate boundary The place where plates meet on the earth's surface and where most of the world's earthquakes occur and volcanoes may be found. *30–31, 90, 113*

Plates Large sections of the earth's crust. *30–31*

Pollution Noise, dirt and other harmful substances produced by people and machines which spoil water, land or the air. *50, 101, 118–119*

Prevailing wind The direction from which the wind usually comes. *6–8, 13, 16–17*

Primary activities Jobs that involve the collecting and using of natural resources, e.g. farming, fishing, mining and forestry. *48, 100, 102, 105, 130–131, 134–135*

Primary goods Raw materials such as minerals, timber and foodstuffs. They are usually of low value. *130, 132–133*

Quality of life A measure of how contented people are with their lives and the environment in which they live and work. *83, 92, 104–105, 119, 122, 128–129*

Raw materials Natural resources that are used to make things. *116–117, 120*

Regenerate Renewing and improving something that has been lost or destroyed. *83–84*

Region An area of land with similar characteristics, e.g. climate, vegetation, political or economic activities. *96–103, 106*

Relief The shape of the land surface and its height above sea-level. *6–7, 13*

Relief rainfall Rain caused by air being forced to rise over hills and mountains. *7–8, 16*

Resources Things that can be useful to people. They may be natural like coal and iron ore or of other value like money and skilled workers. *128*

Richter scale A scale used to measure the strength of an earthquake. *38–39*

'Ring of Fire' A circle of active volcanoes found around the edge of the Pacific Ocean. *28, 30*

Satellite image A photo taken from a satellite orbiting in space and sent back to earth. The images can show either true or false colours. *95, 112, 118*

Scrub Small stunted trees and bushes. *18–19, 25, 94–95, 98*

Secondary activities Where natural resources are turned, or manufactured, into goods that we can use, e.g. cars and computers. *48, 100, 102, 123, 130–131*

Seismograph An instrument used to measure the strength of an earthquake. *29, 40*

Self-help scheme Where local people are involved in improving conditions for themselves, e.g. building better housing or reducing soil erosion. *23*

Sense of identity A feeling of being part of a group, through shared customs and a way of life. *93*

Service activities Jobs that provide a service for people, e.g. teaching and nursing. *48, 100, 102, 123, 130–131*

Silt A fine, fertile soil left behind when a river floods. Also called alluvium. *98, 100*

Site The place where a settlement, factory or stadium is located. *80–81, 117*

Social indicators Development indicators that help measure standards of living, e.g. school attendance and literacy rates. *122–123*

Soil Loose material on the earth's surface in which plants grow and organisms live. *10, 20, 98, 113*

Soil erosion The removal and loss of soil mainly by wind, rain and running water. *20–23, 98, 107*

Species Groups of plants and animals. *15*

Standard of living How well-off a person or a country is. *49, 92, 100, 102, 104–105, 119, 122, 128, 130, 134, 136, 140*

Subsistence farming Growing just enough food for your own needs with nothing left to sell. *102*

Sustainable development A way of improving people's standard of living and quality of life without wasting resources or harming the environment. *118–119, 136–137, 140–141*

Sweatshops Factories where people have to work long hours for very little money. *74–75*

Terracing A flat shelf cut into the hillside and used for growing crops, e.g. rice. *23, 94–95*

Tertiary activities Jobs that provide a service for people, e.g. teaching and nursing. *48, 100, 102, 123, 130–131*

Theme park A purpose-built tourist resort where the attractions are related. *60–61*

Tourism When people travel to places for recreation and leisure. *23, 46–67, 94–97*

Trade The movement and sale of goods between countries. *89, 104, 120–121, 132–133, 139*

Trade deficit When a country spends more on its imports than it earns from its exports. *121*

Trade surplus When a country earns more from its exports than it spends on its imports. *121, 123*

Transnational corporations Large companies with offices and factories across the world. *71–75, 82*

Tropical rainforest Tall, dense forest found in hot, wet climates. *10–11, 14–15, 25*

Vent An opening in the earth's crust through which material is forced upwards during a volcanic eruption. *32, 44*

Volcanic bombs Large rock fragments thrown out by an erupting volcano. *32, 34*

Volcano A cone-shaped mountain or hill often made up of lava and ash. *26–35, 44–45, 90, 94, 98, 110, 112–113*

Weather The day-to-day state of the atmosphere. It includes temperature, rainfall, sunshine and wind. *6, 12, 106*

Zones of activity Areas, usually along plate boundaries, where earthquakes and volcanoes are common. *29*